Modern Well Test Analysis

A Computer-Aided Approach

Roland N. Horne

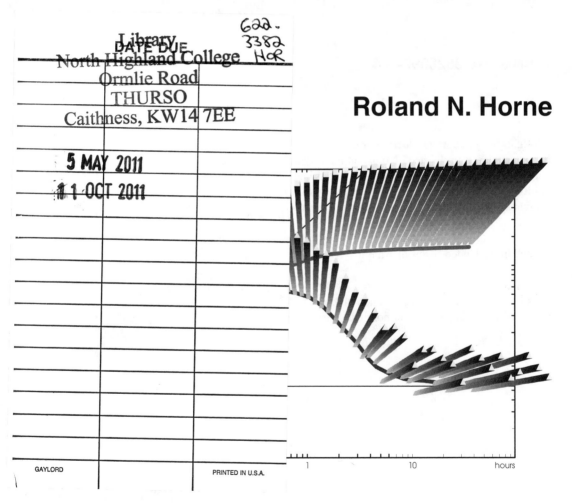

Second Edition

COPYRIGHT

Petroway, Inc.
926 Bautista Court
Palo Alto, CA 94303-4046
(650)494-2037
http://www.petroway.com

Acknowledgements

This book brings together traditional material from earlier sources, together with new material prepared specifically for this work. In particular, I have made use of personal notes of Henry J. Ramey, Jr., and Matt Mavor, as well as the many references listed in the text. The work of these earlier authors who have collected material on well test analysis is gratefully acknowledged.

As a book about computer-aided techniques, this manuscript was produced entirely electronically, including all of the diagrams. None of the figures were drawn by hand, and no paste-up was required. Typesetting and sketch diagrams used Microsoft® Word for Windows™ 6.0 and MS Draw. The layout and indexing macros from Doc-To-Help® by WexTech Systems, Inc. were also used. Most of the well test diagrams were produced directly by the software packages Automate-II and Automate for Windows™ by Munro Garrett International, Inc. Some of the figures were created using AutoSketch by Autodesk, Inc., HiJaak® Draw™ by Inset Systems, Inc., Microsoft Excel 4.0 and Microsoft Powerpoint 3.0. The engineering calculations and typesetting were performed on an Intel Pentium P90 computer running under Microsoft® Windows™ for Workgroups 3.11, and draft printed on an Apple LaserWriter Pro 630. Final printing used a Linotronic 530 printer at 2540 dpi.

Much of the original edition of this manuscript was written while I was a guest of the Petroleum Engineering Department at Heriot-Watt University in Edinburgh, Scotland, and I am indebted to Prof. Jim Peden and his faculty and staff for their hospitality and support during my visit. The first draft of this material was originally prepared for a short course organized in Jakarta, Indonesia, by Jamin Djuang of P.T. Loka Datamas Indah in February 1990. Subsequent to this first "test flight", corrections and expansions brought the manuscript to its first edition. After four years of using this book in industrial short courses and in well testing courses at Stanford University, I decided that it was need of further "modernization" if the title was to remain "Modern Well Test Analysis" -- hence this second edition. For suggestions and corrections, I would like to thank Jim Mallinson of Mobil, Dr. Ueda of Arabian Oil Company and Richard Hughes of Stanford University -- any remaining errors are my own. Based on its origins as course material, this book is intended as a tutorial on well testing, rather than a full textbook or comprehensive reference.

The first edition of this book was dedicated to my friend and mentor, Hank Ramey. Regrettably, Prof. Ramey passed away in November 1993. Although he is no longer with us, the methods of well test analysis he developed live on -- and so he still deserves the title of "the father of modern well test analysis". I dedicate this second edition of the book to his memory.

Roland N. Horne
Palo Alto, California
May 1995

Contents

6. DESIGNING WELL TESTS   127

7. ADVANCED TOPICS   133

8. WORKED EXAMPLES   145

1. WELL TEST OBJECTIVES

1.1 Introduction

During a well test, the response of a reservoir to changing production (or injection) conditions is monitored. Since the response is, to a greater or lesser degree, characteristic of the properties of the reservoir, it is possible in many cases to infer reservoir properties from the response. Well test interpretation is therefore an inverse problem in that model parameters are inferred by analyzing model response to a given input.

In most cases of well testing, the reservoir response that is measured is the **pressure response**. Hence in many cases **well test analysis** is synonymous with **pressure transient analysis**. The pressure transient is due to changes in production or injection of fluids, hence we treat the flow rate transient as **input** and the pressure transient as **output**.

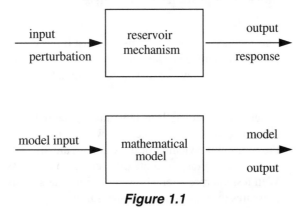

Figure 1.1

In well test interpretation, we use a mathematical model to relate pressure response (output) to flow rate history (input). By specifying that the flow rate history input in the model be the same as that in the field, we can infer that the model parameters and the reservoir parameters are the same if the model pressure output is the same as the measured reservoir pressure output. Clearly, there can be major difficulties involved in this process, since the model may act like the actual reservoir even though the physical assumptions are entirely invalid. This ambiguity is inherent in all inverse problems, including many others used in reservoir engineering (e.g., history matching in simulation, decline curve analysis, material balance). However, the dangers can be minimized by careful specification of the well test in such a way that the response is most characteristic of the reservoir parameters under investigation. Thus in most cases, the design and the interpretation of a well test is dependent on its objectives.

The objectives of a well test usually fall into three major categories:

(a) reservoir evaluation,

(b) reservoir management, and

(c) reservoir description.

1.2 Reservoir Evaluation

To reach a decision as how best to produce a given reservoir (or even whether it is worthwhile to spend the money to produce it at all) we need to know its **deliverability**, **properties**, and **size**. Thus we will attempt to determine the reservoir **conductivity** (kh, or **permeability-thickness** product), **initial reservoir pressure**, and the reservoir limits (or boundaries). At the same time, we will sample the fluids so that their physical properties can be measured in the laboratory. Also, we will examine the near wellbore condition in order to evaluate whether the well productivity is governed by wellbore effects (such as **skin** and **storage**) or by the reservoir at large.

The conductivity (kh) governs how fast fluids can flow to the well. Hence it is a parameter that we need to know to design well spacing and number of wells. If conductivity is low, we may need to evaluate the cost-effectiveness of stimulation.

Reservoir pressure tells us how much potential energy the reservoir contains (or has left) and enables us to forecast how long the reservoir production can be sustained. Pressures in the vicinity of the wellbore are affected by drilling and production processes, and may be quite different from the pressure and the reservoir at large. Well test interpretation allows us to infer those distant pressures from the local pressures that can actually be measured.

Analysis of reservoir limits enables us to determine how much reservoir fluid is present (be it oil, gas, water, steam or any other) and to estimate whether the reservoir boundaries are closed or open (with aquifer support, or a free surface).

1.3 Reservoir Management

During the life of a reservoir, we wish to monitor performance and well condition. It is useful to monitor changes in average reservoir pressure so that we can refine our forecasts of future reservoir performance. By monitoring the condition of the wells, it is possible to identify candidates for workover or stimulation. In special circumstances, it may also be possible to track the movement of fluid fronts within the reservoir, such as may be seen in water flooding or in-situ combustion. Knowledge of the front location can allow us to evaluate the effectiveness of the displacement process and to forecast its subsequent performance.

1.4 Reservoir Description

Geological formations hosting oil, gas, water and geothermal reservoirs are complex, and may contain different rock types, stratigraphic interfaces, faults, barriers and fluid fronts. Some of these features may influence the pressure transient behavior to a measurable extent, and most will affect the reservoir performance. To the extent that it is possible, the use of well test analysis for the purpose of reservoir description will be an aid to the forecasting of reservoir performance. In addition, characterization of the reservoir can be useful in developing the production plan.

Examples of the use of well test analysis for reservoir description can be found in Britt et al. (1989), Myers et al. (1980), Lee (1982), Currier (1988) and Roest et al. (1986). However, it is important to acknowledge that there is a limit to the level of detail that can be achieved in a reservoir description. This is because pressure transmission is an inherently *diffusive* process, and hence is governed largely by average conditions rather than by local heterogeneities. For example, Grader and Horne (1988) showed that it is possible to have a "hole" in the reservoir that is as large as half the distance between a production well and an observation well, without that "hole" making any discernible difference in an interference test. This observation appears discouraging at first, however it underlines the overall usefulness of well test analysis -- well tests can be interpreted to estimate bulk reservoir properties *because* they are insensitive to most local scale heterogeneities.

1.5 Decline Curve Analysis

The discussions above have referred to pressure transient analysis, in which the pressure transient is considered to be the response of a system to a specific flow rate history. It should be clear however that it is equally valid to consider a flow rate response to a specific pressure history. This case, in which well flowing pressure is usually treated as constant and production rate declines, is commonly known as **decline curve analysis**.

Fundamentally, there is no difference between pressure transient analysis and decline curve analysis, however, there are practical considerations that usually separate the two applications. Since flow rate is the easier of the two functions to control in a short term test, pressure transient tests (such as drawdown, buildup and interference tests) are usually conducted over only a few hours or days. Hence pressure transient tests are usually used to diagnose near wellbore conditions, such as kh, storage and skin. During long term production, pressure is often controlled by production equipment requirements, and production rates are monitored in the long term (over months and years) for decline curve analysis. Hence decline curve analysis is more diagnostic of long term effects, such as reservoir volume.

In general terms, both flow rate and pressure are interdependent, and both are governed by reservoir characteristics. Thus pressure transient analysis and decline curve analysis are specific examples of the same process, although traditionally each has been developed somewhat differently. In this book, the distinction between them will be removed. Any time that flow rate and pressure both are measured it is possible to specify one and match the other. Because of the diffusive nature of pressure transmission mentioned earlier, it is often easier to match pressures than it is to match flow rates.

1.6 Types of Tests

In some cases, the type of test performed is governed by the test objectives. In other cases the choice is governed by practical limitations or expediencies. For the purpose of later discussion, the various types of test will be defined in this section.

1.6.1 Drawdown Test

In a drawdown test, a well that is static, stable and shut-in is opened to flow. For the purposes of traditional analysis, the flow rate is supposed to be constant (Figure 1.2).

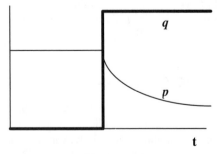

Figure 1.2

Many of the traditional analysis techniques are derived using the drawdown test as a basis (including many of the derivations in Section 2). However, in practice, a drawdown test may be rather difficult to achieve under the intended conditions. In particular:

(a) it is difficult to make the well flow at constant rate, even after it has (more-or-less) stabilized, and

(b) the well condition may not initially be either static or stable, especially if it was recently drilled or had been flowed previously.

On the other hand, drawdown testing is a good method of reservoir limit testing, since the time required to observe a boundary response is long, and operating fluctuations in flow rate become less significant over such long times.

1.6.2 Buildup Test

In a buildup test, a well which is already flowing (ideally at constant rate) is shut in, and the downhole pressure measured as the pressure builds up (Fig. 1.3). Analysis of a buildup test often requires only slight modification of the techniques used to interpret constant rate drawdown test. The practical advantage of a buildup test is that the constant flow rate condition is more easily achieved (since the flow rate is zero).

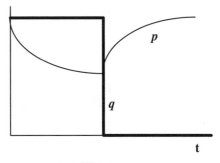

Figure 1.3

Buildup tests also have disadvantages:

(a) It may be difficult to achieve the constant rate production prior to the shut in. In particular, it may be necessary to close the well briefly to run the pressure tool into the hole.

(b) Production is lost while the well is shut in.

1.6.3 Injection Test

An injection test is conceptually identical to a drawdown test, except that flow is into the well rather than out of it (Fig. 1.4).

Injection rates can often be controlled more easily than production rates, however analysis of the test results can be complicated by multiphase effects unless the injected fluid is the same as the original reservoir fluid.

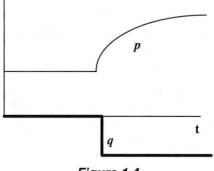

Figure 1.4

1.6.4 Falloff Test

A falloff test measures the pressure decline subsequent to the closure of an injection (Fig. 1.5). It is conceptually identical to a buildup test.

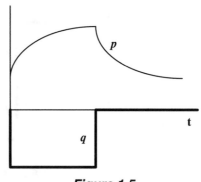

Figure 1.5

As with injection tests, falloff test interpretation is more difficult if the injected fluid is different from the original reservoir fluid.

1.6.5 Interference Test

In an interference test, one well is produced and pressure is observed in a different well (or wells). An interference test monitors pressure changes out in the reservoir, at a distance from the original producing well. Thus an interference test may be useful to characterize reservoir properties over a greater length scale than single-well tests. Pressure changes at a distance from the producer are very much smaller than in the producing well itself, so interference tests require sensitive pressure recorders and may take a long time to carry out. Interference tests can be used regardless of the type of pressure change induced at the active well (drawdown, buildup, injection or falloff).

1.6.6 Drill Stem Test (DST)

A drill stem test is a test which uses a special tool mounted on the end of the drill string. It is a test commonly used to test a newly drilled well, since it can only be carried out while a rig is over the hole. In a DST, the well is opened to flow by a valve at the base of the test tool, and reservoir fluid flows up the drill string (which is usually empty to start with). A common test sequence is to produce, shut in, produce again and shut in again. Drill stem tests can be quite short, since the positive closure of the downhole valve avoids wellbore storage effects (described later). Analysis of the DST requires special techniques, since the flow rate is not constant as the fluid level rises in the drill string. Complications may also arise due to momentum and friction effects, and the fact that the well condition is affected by recent drilling and completion operations may influence the results.

1.7 References

Britt, L.K., Jones, J.R., Pardini, R.E., and Plum, G.L.: "Development of a Reservoir Description Through Interference Testing of the Clayton Field, Prairie du Chien Formation", paper SPE 19846, *Proceedings* SPE 64th Annual Fall Technical Conference and Exhibition, San Antonio, TX, Oct. 8-11, (1989), 787-800.

Currier, B.H.: "Lisburne Reservoir Limited Drainage Test: A Pilot Test Case History", paper SPE 18277, presented at SPE 63rd Annual Fall Technical Conference and Exhibition, Houston, TX, Oct. 2-5, (1988).

Grader, A., and Horne, R.N.: "Interference testing: Detecting an Impermeable or Compressible Sub-Region," *SPE Formation Evaluation*, (1988), 428-437.

Lee, B.O.: "Evaluation of Devonian Shale Reservoirs Using Multiwell Pressure Transient Testing Data," paper SPE/DOE 10838, presented at the SPE/DOE Unconventional Gas Recovery Symposium, Pittsburgh, PA, May 16-18, (1982).

Myers, G.A., Johnson, R.D., and Mainwaring, J.R.: "Simulation of Prudhoe Bay Field Interference Test," paper SPE 9456, presented at SPE 55th Annual Fall Technical Conference and Exhibition, Dallas, TX, Sept. 21-24, (1980).

Roest, J.A., Jolly, D.C., and Rogriguez, R.A.: "Pulse Testing Reveals Poor Lateral and Vertical Continuity in a Reservoir Consisting of Distributary Channel Sands", paper SPE 15613, presented at SPE 61st Annual Fall Technical Conference and Exhibition, New Orleans, LA, Oct. 5-8 (1986).

2. WELL TEST CONCEPTS

2.1 Basics of Reservoir Models

In the process of testing a well, we provide an input **impulse** (usually a change in flow rate) and we measure the **response** (usually a change in pressure). The reservoir response is governed by parameters such as permeability, skin effect, storage coefficient, distance to boundaries, fracture properties, dual porosity coefficients, etc. Based on an understanding of the reservoir physics, we develop a mathematical model of the dependence of the response on these reservoir parameters. Then by matching the **model response** to the measured **reservoir response** we infer that the model parameters take the same values of the reservoir parameters. This process is illustrated in Fig. 2.1.

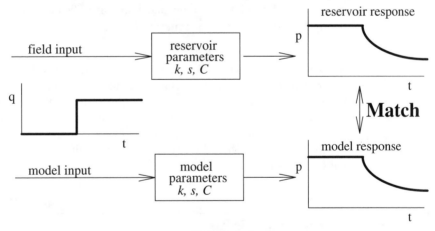

Figure 2.1

The mathematical model can be either analytical or numerical (i.e., a reservoir simulator), but is usually analytical. The remainder of this section will discuss the basis of these analytical reservoir models.

The mathematical equation governing pressure transmission in a porous medium filled by slightly compressible fluid is given by (in cylindrical coordinates):

$$\frac{\partial^2 p}{\partial r^2} + \frac{1}{r}\frac{\partial p}{\partial r} + \frac{k_\theta}{k_r}\frac{1}{r^2}\frac{\partial^2 p}{\partial \theta^2} + \frac{k_z}{k_r}\frac{\partial^2 p}{\partial z^2} = \frac{\phi\mu c_t}{k_r}\frac{\partial p}{\partial t}$$

(2.1)

Assumptions inherent in this equation are:

(a) Darcy's Law applies,

(b) Porosity, permeabilities, viscosity and compressibility are constant,

(c) Fluid compressibility is small (this equation is not usually valid for gases),

(d) Pressure gradients in the reservoir are small (this may not be true in high rate wells or for gases),

(e) Flow is single phase,

(f) Gravity and thermal effects are negligible.

If permeability is isotropic, and only radial and vertical flow are considered, then this equation reduces to:

$$\frac{\partial^2 p}{\partial r^2} + \frac{1}{r}\frac{\partial p}{\partial r} + \frac{\partial^2 p}{\partial z^2} = \frac{\phi\mu c_t}{k_r}\frac{\partial p}{\partial t}$$

(2.2)

This equation is recognizable as the **diffusion equation**, which appears in many fields of science and engineering. This underlines the importance of the diffusion process in well test interpretation, and emphasizes the underlying significance of the **hydraulic diffusivity** parameter $k/\phi\mu c_t$.

For the purposes of this book (and for the application of well test interpretation in general), it is not necessary for us to understand the process of solution of the pressure transmission equation (Eq. 2.2). Solutions to this equation have been developed for a wide variety of specific cases, covering many reservoir configurations. These specific reservoir solutions are the models that we will use to match reservoir behavior, thereby inferring reservoir parameters that we do not know in advance.

2.2 Dimensionless Variables

Well test analysis often makes use of dimensionless variables. The importance of dimensionless variables is that they simplify the reservoir models by embodying the reservoir parameters (such as k), thereby reducing the total number of unknowns. They have the additional advantage of providing model solutions that are independent of any particular unit system. It is an inherent assumption in the definition that permeability, viscosity, compressibility, porosity, formation volume factor and thickness are all constant.

The **dimensionless pressure** p_D is defined (in oilfield units) as:

$$p_D = \frac{kh}{141.2qB\mu}\left(p_i - p_{wf}\right)$$

$$(2.3)$$

where:

k	=	permeability (md)
h	=	thickness (feet)
p_i	=	initial reservoir pressure (psi)
p_{wf}	=	well flowing pressure (psi)
q	=	production rate (STB/d)
B	=	formation volume factor (res vol/std vol)
μ	=	viscosity (cp)

In a consistent unit set, p_D is defined as:

$$p_D = \frac{2\pi kh}{qB\mu}\left(p_i - p_{wf}\right)$$

$$(2.3a)$$

The dimensionless time t_D is defined (in oilfield units) as:

$$t_D = \frac{0.000264kt}{\phi\mu c_t r_w^2}$$

$$(2.4)$$

where:

t	=	time (hours)
ϕ	=	porosity (pore volume/bulk volume)
c_t	=	total system compressibility (/psi)
r_w	=	wellbore radius (ft)

In a consistent unit set, t_D is defined as:

$$t_D = \frac{kt}{\phi\mu c_t r_w^2}$$

$$(2.4a)$$

This is only one form of the dimensionless time. Another definition in common usage is t_{DA}, the dimensionless time based upon reservoir area:

$$t_{DA} = \frac{0.000264kt}{\phi\mu c_t A}$$

$$(2.5)$$

where:

A	=	reservoir area = πr_e^2
r_e	=	reservoir radius (ft)

Clearly there is a direct relationship between t_D and t_{DA}:

$$t_D = t_{DA} \frac{A}{r_w^2} = t_{DA} \pi \frac{r_e^2}{r_w^2} \qquad (2.6)$$

We can also define a dimensionless radius, r_D, as:

$$r_D = \frac{r}{r_w} \qquad (2.7)$$

This definition is independent of any particular set of units.

2.3 The Skin Effect

Pressure transmission does not take place uniformly throughout the reservoir, since it is affected by local heterogeneities. For the most part, these do not affect the pressure change within the well, except those reservoir heterogeneities in the immediate vicinity of the wellbore. In particular, there is often a zone surrounding the well which is invaded by mud filtrate or cement during the drilling or completion of the well -- this zone may have a lower permeability than the reservoir at large, and thereby acts as a "skin" around the wellbore, causing higher pressure drop. This is shown in Fig. 2.2.

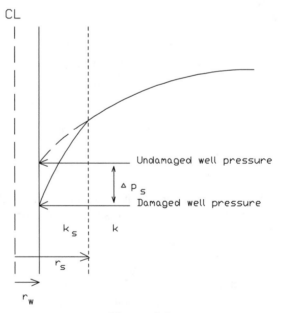

Figure 2.2

The pressure drop across the skin Δp_s is the difference between the actual pressure in the well when it is flowing, and the pressure that would have been seen if the well were undamaged. The **skin factor** is a variable used to quantify the magnitude of the skin effect. From its definition, we can see that the skin factor is actually a dimensionless pressure. The skin factor s is defined (in oilfield units) as:

$$s = \frac{kh}{141.2qB\mu} \Delta p_s$$

(2.8)

In consistent units, the skin effect would be defined:

$$s = \frac{2\pi kh}{qB\mu} \Delta p_s$$

(2.9)

If we imagine that the skin effect is due to a damaged zone of radius r_s and reduced permeability k_s, then the skin effect can be calculated from:

$$s = \left(\frac{k}{k_s} - 1\right) \ln \frac{r_s}{r_w}$$

(2.10)

We can also describe the skin effect in terms of an **effective wellbore radius**. This is the smaller radius that the well appears to have due to the reduction in flow caused by the skin effect. This effective radius is given by:

$$r_{weff} = r_w e^{-s}$$

(2.11)

It can be seen from Eq. 2.10 that if the skin zone permeability k_s is higher than that of the reservoir (as can happen due to acidization or stimulation), then the skin effect can be **negative**. In the case of negative skin, the effective wellbore radius, given by Eq. 2.11, will be greater than the actual radius. The pressure distribution in this case would appear as in Fig. 2.3.

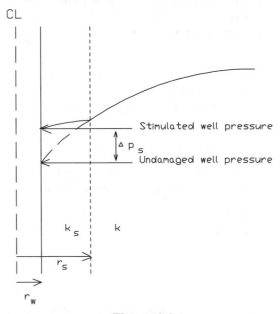

Figure 2.3

Clearly there is a limit as to how negative a negative skin effect can be, and in practice skin factors less than -5 are rarely seen. This is not the case for positive skin effects, which can have any magnitude (although they are rarely greater than 20).

Example 2.1

A 12 inch diameter hole has a damaged region 48 inches deep. The permeability in this region is one tenth that of the undamaged region. The skin effect can be estimated as follows:

$r_w = 0.5$ feet

$r_s = r_w + 4 = 4.5$ feet

from Eq. 2.10:

$s = (k/k_s - 1) \ln r_s/r_w = (10 - 1) \ln 4.5/0.5$

$\underline{s = 19.78}$

If, on the other hand, the 48 inch deep zone were stimulated instead of damaged, and the permeability enhancement was a factor of 10, then the skin factor would be:

$s = (0.1 - 1) \ln 4.5/0.5$

$\underline{s = -1.98}$

From these two numerical examples, it can be seen that the numerical values of positive and negative skins are quite different. A tenfold *decrease* in permeability gives rise to a positive skin factor of about 20, whereas a tenfold *increase* in permeability produces a negative skin factor of only -2.

It is interesting also to look at the effective wellbore radii for the two cases just described. For the positive skin factor of 19.78, the effective wellbore radius will be, from Eq. 2.11:

$r_{weff} = 0.5 \ e^{-19.78}$

$r_{weff} = 1.28 \times 10^{-9}$ feet

From this we can see that a skin of 19.78 is very large indeed! For the negative skin factor of -1.98, the effective wellbore radius will be:

$r_{weff} = 0.5 \ e^{1.98}$

$r_{weff} = 3.62$ feet

2.3.1 Flow Efficiency

A term sometimes used to describe the wellbore damage is **flow efficiency**, the ratio of the theoretical pressure drop if no skin had been present to the actual pressure drop measured during the test. The flow efficiency parameter can be used to calculate the flow rate that could be achieved if the wellbore damage were removed (by stimulation) since it is also the ratio of the ideal (zero skin) flow rate to the actual flow rate.

$$FE = \frac{\Delta p_{(zero\ skin)}}{\Delta p_{(actual)}} = \frac{q_{(actual)}}{q_{(zero\ skin)}}$$

$$(2.12)$$

In practice, flow efficiency is time dependent, not because the well damage is changing, but simply due to its mathematical definition. Thus it is not as definitive as parameter as skin factor, which is a constant for all time.

2.3.2 Partial Penetration Skin

Skin effect is not always due just to wellbore damage. If a well has limited entry, or only partially penetrates the formation, then flow cannot enter the well over the entire producing interval and the well will experience a larger pressure drop for a given flow rate than a well that fully penetrates the formation. This geometric effect gives rise to the **partial penetration skin** effect. It is often useful to estimate the size of the partial penetration skin factor, since it can be subtracted from the apparent overall skin to determine whether the well is actually damaged. Kuchuk and Kirwan (1987) provided an equation for the computation of the partial penetration skin factor:

$$s_{pp} = \frac{2}{\pi b} \sum_{n=1}^{\infty} \frac{1}{n} \sin(n\pi b) \cos(n\pi b z_D^*) K_0\left(\frac{n\pi b}{h_{wD}}\right)$$

$$(2.13)$$

where b is the penetration ration (h_w/h), k_h and k_v are the horizontal and vertical permeabilities respectively, and:

$$h_{wD} = \frac{h_w}{r_w} \sqrt{\frac{k_h}{k_v}}$$

$$(2.14)$$

The effective average pressure point z_D^* was computed by Gringarten and Ramey (1975), and can be found from Fig. 2.4 or from the approximate polynomial expression:

$$z_D^* = 0.9069 - 0.05499 \log h_{wD} + 0.003745(\log h_{wD})^2$$

$$(2.15)$$

Using Eqs. 2.13 and 2.15, the partial penetration skin effect can be computed. Alternatively, an approximate value sufficient for most practical analyses can be read from the graphical representation in Fig. 2.5.

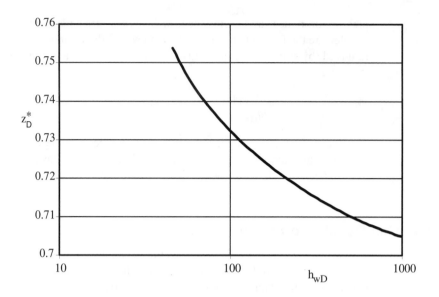

Figure 2.4

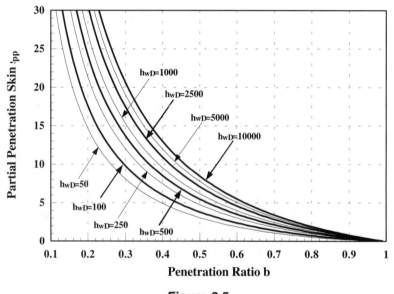

Figure 2.5

2.4 Wellbore Storage

We have understood that, in most cases, well test analysis is the interpretation of the pressure response of the reservoir to a given change in the rate (from zero to a constant value for a drawdown test, or from a constant value to zero for a buildup test). However, for many well tests, the only means of controlling the flow rate is at the wellhead valve or flow line. Hence although the well may produce at constant rate at the wellhead, the flow transient within the wellbore itself may mean that the flow rate from the reservoir into the wellbore (the "sand face" flow rate, q_{sf}) may not be constant at all. This effect is due to **wellbore storage**.

Wellbore storage effect can be caused in several ways, but there are two common means. One is storage by **fluid expansion**, the other is storage by **changing liquid level**.

Consider the case of a drawdown test. When the well is first open to flow, the pressure in the wellbore drops. This drop causes an expansion of the wellbore fluid, and thus the first production is not fluid from the reservoir but is fluid that had been stored in the wellbore volume. As the fluid expands, the wellbore is progressively emptied, until the wellbore system can give up no more fluid, and it is the wellbore itself which provides most of the flow during this period. This is wellbore storage due to fluid expansion.

The second common kind of wellbore storage is due to a changing liquid level. This is easily envisaged in the case of a completion consisting of a tubing string without a packer, as in Fig. 2.6.

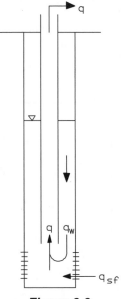

Figure 2.6

When the well is open to flow during a drawdown test, the reduction in pressure causes the liquid level in the annulus to fall. The liquid extracted from the annulus joins that from the reservoir and makes up a proportion of the total flow from the well. The falling liquid level is generally able to supply much more fluid than is possible simply from expansion of the fluid alone, thus wellbore storage effects are usually much more prominent in this type of completion.

The **wellbore storage coefficient**, C, is a parameter used to quantify the effect. C is the volume of fluid that the wellbore itself will produce due to a unit drop in pressure:

$$C = \frac{V}{\Delta p} \tag{2.16}$$

where V is the volume produced, and Δp is the pressure drop. C has units STB/psi (or sometimes MCF/psi in the case of gas wells). It is also common to use a **dimensionless wellbore storage coefficient**, C_D, defined as

$$C_D = \frac{5.615C}{2\pi\phi c_t h r_w^2} \tag{2.17}$$

where C is in STB/psi.

Assuming the fluid is of constant density, conservation of mass requires that the total flow rate q be equal to the flow of fluid from the reservoir (q_{sf}) added to that which flows from the well itself (q_w):

$$q = q_{sf} + q_w \tag{2.18}$$

Thus the fraction of the total flow that originates from the reservoir is given by:

$$\frac{q_{sf}}{q} = 1 - \frac{q_w}{q} \tag{2.19}$$

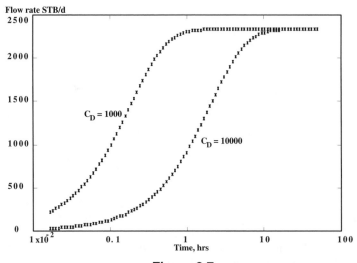

Figure 2.7

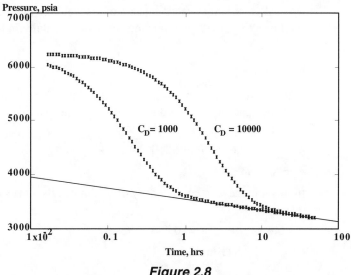

Figure 2.8

The overall effect of wellbore storage can be seen in Fig. 2.7. At early time the ratio q_{sf}/q is close to zero, as all the fluid produced at the wellhead originates in the wellbore. As time goes on, the wellbore storage is depleted, and eventually the reservoir produces all the fluid (as q_{sf}/q tends to one). The corresponding pressure transients due to the wellbore storage effects are seen in Fig. 2.8. It is important to recognize that, as a consequence of the wellbore storage effect, the early transient response during a well test is not characteristic of the reservoir, only of the wellbore. This means that a well test must be long enough that the wellbore storage effect is over and fluid is flowing into the wellbore from the reservoir. As we will see later, we can also overcome the problem of wellbore storage by specifically measuring the sandface flow rate q_{sf} down hole.

The value of the wellbore storage coefficient can be estimated based upon the configuration of the completion. For a **fluid expansion** storage coefficient,

$$C = c_w V_w$$

(2.20)

where V_w is the volume of the wellbore, and c_w is the compressibility of the fluid in the wellbore. In principle, the wellbore compressibility includes the volume changes in the tubing and casing, however, these are usually small. Nonetheless, the compressibility is different from c_t the total reservoir compressibility, since c_t includes the rock compressibility and will be under different pressure, temperature and saturation conditions than the wellbore.

For a **falling liquid level** storage coefficient:

$$C = \frac{144 A_w}{\rho} ft^3 / psi$$

(2.21)

where A_w is the cross-sectional area of the wellbore in the region where the liquid level is falling (in ft^2) and ρ is the density of the fluid (in lbm/ft^3).

Example 2.2

A 1000 ft deep pumping well has a 2 inch OD tubing inside a 7-5/8 inch ID casing. Without a packer, the liquid level will be pumped down over the annular space. This space has a cross sectional area of:

$$A_w = \pi \, [\, (7.675/2)^2 - (2/2)^2] \, 1/144$$

$$A_w = 0.2995 \ \text{ft}^2$$

If the oil in the well has a density of 58 lbm/ft^3, then the storage coefficient will be (using Eq. 2.21):

$$C = 144 \times 0.2995/58 = 0.743 \ \text{ft}^3/\text{psi}.$$

$$C = 0.132 \ \text{STB/psi}.$$

If the same well were filled with a gas (of compressibility 2×10^{-4} /psi^{-1}), then instead of the falling liquid level effect we would see a fluid expansion storage effect (using Eq. 2.20):

$$C = 2 \times 10^{-4} \, V_w$$

where the wellbore volume is:

$$V_w = \pi \, (7.625/2)^2 \, (1000/144) \ \text{ft}^3$$

So, $C = 2 \times 10^{-4} \, \pi \, (7.625/2)^2 \, 1000/144 = 0.0634 \ \text{ft}^3/\text{psi}.$

$$C = 0.011 \ \text{STB/psi}.$$

Notice that even though the gas is much more compressible, the falling liquid level storage coefficient is very much larger than the fluid expansion storage coefficient.

Wellbore storage is a major nuisance to well test interpretation, since it disguises the reservoir response until late in the test. One way to overcome this problem is to measure the flow rates downhole instead of at the surface. Such flow rate measurements have become more common during the 1980's, but are still by no means standard. Measuring flow rates adds to the cost of the test, and is difficult in multiphase, inclined and pumping wells. Reservoir engineers still need to be prepared to analyze well test results in the presence of wellbore storage.

From material balance, the pressure in the wellbore is directly proportional to time during the wellbore storage dominated period of the test:

$$p_D = \frac{t_D}{C_D} \tag{2.22}$$

On a log-log plot of pressure drop versus time, this gives a characteristic straight line of unit slope (Fig. 2.9).

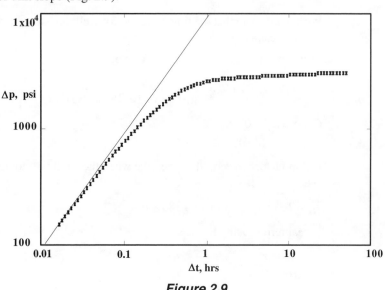

Figure 2.9

The unit slope straight line response continues up to a time given approximately by:

$$t_D = C_D(0.041 + 0.02s) \tag{2.23}$$

provided that the skin factor s is positive. However, the storage effect is not over at this time, as there is a period (roughly one and a half log cycles long) during which the response undergoes a transition between wellbore response and reservoir response. Thus the reservoir response does not begin until a time:

$$t_D = C_D(60 + 3.5s) \tag{2.24}$$

During the design of a test, care should be taken to ensure that the test is at least this long (and usually very much longer, even if nonlinear regression techniques are to be used for the interpretation).

Example 2.3

The pumping well in Example 2.2 produces from a reservoir with the following properties:

Permeability $k = 20$ md

Thickness $h = 5$ feet

Porosity $\phi = 20\%$

Compressibility $c_t = 5 \times 10^{-6}$ /psi

Wellbore radius $r_w = 0.32$ feet

Viscosity $\mu = 2$ cp

We can estimate that the time required for wellbore storage effects (and transition) to disappear as follows:

(from Example 2.2), $C = 0.743$ ft^3 /psi.

(from Eq. 2.17) $C_D = \dfrac{0.743}{2\pi(0.2)(5\times10^{-6})(5)(0.32)^2}$

$C_D = 2.31 \times 10^5$

If skin factor s is zero, then the time when the wellbore storage influenced response ends is:

$$t_D = 2.31 \times 10^5 (60 + 0) = 1.386 \times 10^7$$

In real terms, using Eq. 2.4:

$$t = \frac{\phi \mu c_t r_w^2}{0.000264k} t_D$$

$$t = \frac{(0.2)(2)(5\times10^{-6})(0.32)^2}{0.000264(20)} 1.386 \times 10^7$$

$t = 537.6$ hrs

This is clearly an impractically long time, and we could not even consider attempting to test this well without a downhole flow rate measurement. If the skin effect is as large as 10, this time would be about 50% larger (from Eq. 2.24).

Based on this example, it can be seen that the time affected by wellbore storage is independent of porosity, wellbore radius and system compressibility. In fact Eqs. 2.24, 2.17 and 2.5 can be combined to give the storage-influenced time as:

$$t = \frac{3385 \ C(60 + 3.5s)}{kh / \mu} \tag{2.25}$$

From this we can see that wells with greater deliverability (related to kh/μ) are influenced less strongly by wellbore storage effects. It is also important to note that the storage effect is independent of the flow rate q.

2.5 Infinite Acting Radial Flow

Once the wellbore storage effects are over, the wellbore pressure transient reflects the pressure transmission out in the reservoir. As time proceeds, the response is characteristic of conditions further and further away from the wellbore. At very late time, the pressure response is affected by the influence of reservoir boundaries, but prior to those late times the pressure response does not "see" the reservoir boundaries, and the reservoir acts as if it were infinite in extent. This intermediate time response, between the early wellbore-dominated response and the late time boundary-dominated response, is known as the **infinite acting period**. Although modern well test interpretation identifies several different types of flow during the infinite acting period, one of the most common and easily identified is **radial flow**. Infinite acting radial flow has been the basis of very many well test interpretation techniques, and it is worthwhile to discuss it in some detail.

In the absence of wellbore storage and skin effects, the pressure transient due to infinite acting radial flow into a line source wellbore producing at constant flow rate is given by:

$$p_D = -\tfrac{1}{2}\mathrm{Ei}\left(-\frac{r_D^2}{4t_D}\right)$$

(2.26)

Here *Ei* represents the **exponential integral function**. This solution is valid throughout the reservoir ($r_D > 1$), including at the wellbore ($r_D = 1$). Thus it can be used for interference tests as well as drawdown and buildup tests. Fig. 2.10 shows the exponential integral solution plotted in semilog coordinates.

From this graphical presentation, it can be seen that the infinite acting radial flow response is directly proportional to the logarithm of time for all but early times. Examination of the solution numerically confirms this to be true. For $t_D/r_D^2 > 10$, the exponential integral solution at $r_D = 1$ can be approximated by:

$$p_{wD} = \tfrac{1}{2}\left(\ln t_D + 0.80907\right) + s$$

(2.27)

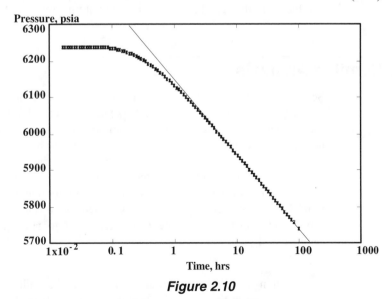

Figure 2.10

Writing this in dimensional variables:

$$p_{wf} = p_i - 162.6 \frac{qB\mu}{kh} \left(\log t + \log \frac{k}{\phi \mu c_t r_w^2} + 0.8686s - 3.2274 \right)$$

$$(2.28)$$

where the natural logarithm (*ln*) has been replaced by a logarithm to base 10 (*log*). From this equation it is seen that a plot of pressure drop against the logarithm of time should contain a straight line with slope:

$$m = 162.6 \frac{qB\mu}{kh}$$

$$(2.29)$$

Hence the recognition of this slope makes it possible to estimate the permeability (*k*) or the permeability-thickness product (*kh*). Many traditional well test interpretation techniques are based on this "semilog approach" and the recognition of the "correct semilog straight line" is a crucial aspect of this type of analysis. We shall examine this further in Section 3.

The skin factor can be estimated from the difference between p_i and the intercept of the straight line. This can be done conveniently by substituting the time 1 hour in Eq. (2.28), and solving for *s*:

$$s = 1.151 \left[\frac{p_i - p_{1hr}}{m} - \log \frac{k}{\phi \mu c_t r_w^2} + 3.2274 \right]$$

$$(2.30)$$

It is important to note that the value of p_{1hr} needs to be taken *from the straight line*, or the extrapolation of it. This is due to the assumption of infinite acting radial flow -- the flow regime at the arbitrary time of 1 hour may not be infinite-acting, and therefore the actual pressure data point at this time is not the correct one to use.

2.6 Semilog Analysis

Semilog analysis is based on the location and interpretation of the semilog straight line response (infinite acting radial flow) described in "2.5 Infinite Acting Radial Flow" on page 22. However, it is very important to note that semilog analysis is not based on the semilog straight line alone, since it is first necessary to determine, based on the duration of the wellbore storage effect, the time at which the semilog straight line begins. Not all well tests will necessarily include an infinite acting radial flow response period, and an apparent straight line on a semilog plot may not in itself be representative of radial flow. Therefore it is always important to begin a semilog analysis by considering the storage effect, to gain confidence in locating the semilog straight line correctly.

As shown in "2.4 Wellbore Storage" on page 16, the wellbore storage shows as a unit slope straight line on a log-log plot of Δp vs. *t*. From Eqs. 2.23 and 2.24 we saw that there is about 1½ log cycles between the end of the unit slope straight line representing wellbore storage and the start of the purely reservoir response (infinite acting radial flow in cases of interest in the context of semilog analysis). This observation gives rise to the **1½ log cycle rule**, providing us with a useful method of identifying the start of the semilog straight line.

Hence the steps involved in a simple semilog analysis are:

(a) Draw a log-log plot of Δp vs. t.

(b) Determine the time at which the unit slope line ends.

(c) Note the time 1½ log cycles ahead of that point. This is the time at which the semilog straight line can be expected to start.

(d) Draw a semilog plot of p vs. t.

(e) Look for the straight line, starting at the suggested time point.

(f) Estimate the permeability k from the slope of the straight line, using Eq. 2.29.

(g) Estimate the skin factor s from the intercept of the line, using the pressure point at t=1 hour on the straight line (not on the data) and Eq. 2.30.

2.6.1 Semilog Analysis -- Illustrative Example

Consider the example drawdown data shown in Fig. 2.11. Following the steps described earlier in "2.6 Semilog Analysis" the first requirement is to identify the wellbore storage effect. The data have been plotted as $p_i - p(t)$ vs. t on log-log coordinates and a unit slope straight line can be seen at early time, ending at about 0.08 hours. Based on the 1½ log cycle rule, we would expect to see the semilog straight line begin at a time t of about 2 hours.

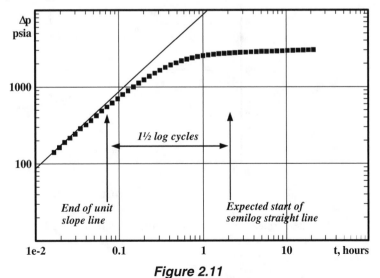

Figure 2.11

Before going ahead to look at the semilog plot, it is possible to make an estimate of the wellbore storage coefficient C based on the unit slope line. In most cases the actual value of the wellbore storage coefficient is not of particular interest, since it is a function of wellbore properties only, however it may be useful to determine the magnitude of the wellbore storage effect for consistency checks of the other parameter estimates. The equation of the unit slope straight line, based on Eq. 2.22 is:

$$(p_i - p) = \frac{0.234}{5.615}\frac{qB}{C}t$$

(2.31)

Any point on the unit slope straight line may be used to obtain the estimate for C, for example the point at $t = 0.0167$ hours, $p = 5867.82$ psia gives an estimate of C of 0.0154 STB/psi. Other parameters of the test required to make the computations in this example are as follows:

B_t (RB/STB)	1.21	r_w (feet)	0.401
μ_o (cp)	0.92	h (feet)	23
c_t (/psi)	8.72×10^{-6}	p_i (psia)	6009
ϕ	0.21	q (STB/d)	2500

Moving on to the semilog plot shown in Fig. 2.12, a semilog straight line can be seen starting at the expected time of about 2 hours. The slope of the line is 255.2 psi/log cycle.

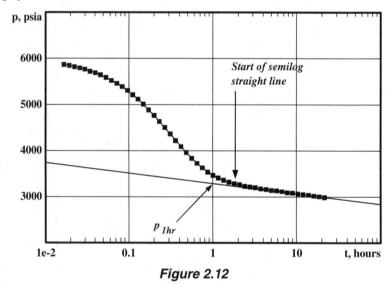

Figure 2.12

The permeability k can be estimated using Eq. 2.29:

$$m = 162.6 \frac{qB\mu}{kh}$$

$$255.2 = 162.6 \frac{(2500)(1.21)(0.92)}{k(23)}$$

Hence the permeability $k = 77.1$ md.

The skin factor may be estimated after picking the pressure point p_{1hr} from the semilog straight line. It is very important to note again that this point lies on the semilog straight line or its extension, and may not lie on the data itself. Picking the pressure from the data set at a time of 1 hour would be incorrect and would give an erroneous estimate of skin factor. The value of from the semilog straight line in Fig. 2.12 is 3330 psia -- the actual pressure data point at 1 hour would have been 3465 psia. Using Eq. 2.30:

$$s = 1.151 \left[\frac{p_i - p_{1hr}}{m} - \log \frac{k}{\phi \mu c_t r_w^2} + 3.2274 \right]$$

$$s = 1.151 \left[\frac{6009 - 3330}{255.2} - \log \frac{77.1}{(0.21)(0.92)(8.72 \times 10^{-6})(0.41)^2} + 3.2274 \right]$$

Hence the skin factor $s = 6.09$.

As a consistency check, the dimensionless time at which the reservoir response begins can be estimated from Eq. 2.25:

$$t = \frac{3385 \ C(60 + 3.5s)}{kh / \mu}$$

$$t = \frac{3385(0.0154)(60 + 3.5 \times 6.09)}{(77.1)(23) / (0.92)} = 2.2 \, hours$$

This is in agreement with the time the data reached the semilog straight line in Fig. 2.12. Thus we have confidence that we have picked the correct semilog straight line.

2.7 Log-Log Type Curves

It should be noted that semilog analysis uses only part of the data (the semilog straight line) to estimate the unknown reservoir parameters. Since the early part of the reservoir response is usually overshadowed by wellbore storage effects, we need to wait until the semilog straight line is visible before a semilog analysis can be performed. Since we know that there is usually about 1½ log cycles of data between the end of wellbore storage and the start of the semilog straight line, we might wonder if there is some way to use this transitional data as well in our analysis. One alternative to semilog analysis that uses the transitional data is **log-log type curve** analysis. Log-log type curve analysis makes use of the dimensionless variables described in "2.2 Dimensionless Variables" on page 10. Since, by definition, dimensionless pressure and time are *linear* functions of actual pressure and time (recall Eqs. 2.3 and 2.4), then the logarithm of actual pressure drop will differ from the logarithm of dimensionless pressure drop *by a constant amount*.

$$\log \Delta p = \log p_D - \log \frac{kh}{141.2qB\mu}$$

(2.32)

similarly

$$\log t = \log t_D - \log \frac{0.000264k}{\phi\mu c_t r_w^2}$$

(2.33)

Hence a graph of log Δp versus log t will have an *identical shape* to a graph of log p_D versus log t_D, although the curve will be shifted by log $kh/141.2qB\mu$ vertically (in pressure) and log $(0.000264\ k/\phi\mu c_t r_w^2)$ horizontally (in time). Matching the two curves will give us estimates of kh from $kh/141.2qB\mu$ (assuming q, B and μ are known) and ϕh (from $(0.000264\ k/\phi\mu c_t r_w^2$, assuming μ, c_t and r_w^2 are known). This process provides a useful method of estimating two very important reservoir parameters, the transmissivity or ability to flow, and the storativity or quantity of fluid contained. Type curves can also be used to estimate other parameters -- this will be illustrated in examples.

2.7.1 Type Curve Analysis -- Illustrative Example

Log-log type curve analysis is one of the most straightforward ways of appreciating the "model matching" process involved in almost all types of well test analysis. By matching the data to the type curve, the reservoir parameters can be estimated. Traditionally, the matching was performed by plotting the data on tracing paper, then sliding the tracing paper over the type curve so that the type curve lines could be seen underneath. Nowadays a type curve match is more commonly done using computer software in which the data is moved "over" the type curve on a computer screen.

Consider the example drawdown data shown in Fig. 2.13. The data have been plotted on log-log coordinates that are the same size as the coordinates of the type curve. By sliding the data over the type curve, the correspondence between Δp (data) and p_D (type curve) can be found. In this example, the last data point in the test was at a time t of 21.6 hours and a pressure p of 2988.93 psia. When the data are lined up to match the underlying type curve as closely as possible, this last data point at $p=2988.93$ psia is found to correspond to a point on the p_D axis of $p_D=13.67$, and a point on the t_D/c_D axis of $t_D/c_D=800$. The other known parameters in the example are as follows:

B_t (RB/STB)	1.21	r_w (feet)	0.401
μ_o (cp)	0.92	h (feet)	23
c_t (/psi)	8.72×10^{-6}	p_i (psia)	6009
ϕ	0.21	q (STB/d)	2500

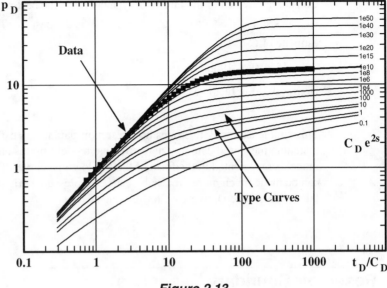

Figure 2.13

Using these values and substituting the pressure match point into Eq. 2.32, we find:

$$\log \Delta p = \log p_D - \log \frac{kh}{141.2qB\mu}$$

$$\log (6009 - 2988.93) = \log 13.67 - \log \frac{k(23)}{141.2(2500)(1.21)(0.92)}$$

Hence, the permeability $k = 78.15$ md.

Substituting the time match point into Eq. 2.33, we find:

$$t_D = \frac{0.000264kt}{\phi\mu c_t r_w^2}$$

$$800C_D = \frac{0.00264(78.15)(21.6)}{(0.21)(0.92)(8.72\times10^{-6})(0.41)^2}$$

Hence, the dimensionless storage coefficient $C_D = 1960$.

From the definition of the dimensionless wellbore storage coefficient (Eq. 2.17):

$$C_D = \frac{5.615C}{2\pi\phi c_t h r_w^2}$$

$$C = 1960\frac{2\pi(0.21)(8.72\times10^{-6})(23)(0.41)^2}{5.615}$$

Hence, the storage coefficient C=0.0155 STB/psi.

Considering which of the family of type curves is the best representation of the data, we can see that the value of $C_D e^{2s} = 5 \times 10^8$ is the best fit. Therefore, using the estimate of C_D:

$$e^{2s} = (5 \times 10^8) / 1960$$

Hence, the skin factor $s=6.22$.

This simple example illustrates how reservoir parameters can be extracted by matching measured data to a mathematical model. In this case, the mathematical model is represented in the form of the log-log type curve. Type curve matching can be made much more precise by using the pressure derivative which is described later in "3.3 The Derivative Plot" on page 67.

2.8 Reservoir Boundary Response

Obviously, reservoirs are not really infinite in extent, thus the infinite acting radial flow period cannot last indefinitely. Eventually the effects of the reservoir boundaries will be felt at the well being tested. The time at which the boundary effect is noticed is dependent on several factors, including the distance to the boundary and the properties of the permeable formation and the fluid that fills it. The two types of reservoir boundary that are most commonly considered are (a) impermeable and (b) constant pressure. An impermeable boundary (also known as a closed boundary) occurs where the reservoir is sealed and no flow occurs. No flow boundaries can also arise due to the interference between wells. A constant pressure boundary rarely occurs exactly in practice, however, in many cases aquifer support, a balanced injection pattern or the presence of a large gas cap can cause an effect that closely approximates a constant pressure boundary.

2.8.1 Closed Boundaries

When a reservoir (or a well's own "drainage region") is closed on all sides, the pressure transient will be transmitted outwards until it reaches all sides, after which the reservoir depletion will enter a state known as **pseudosteady state**. In this state, the pressure in the reservoir will decline at the same rate everywhere in the reservoir (or drainage region). Thus a pseudosteady state is not at all steady, and corresponds to the kind of pressure response that would be seen in a closed tank from which fluid was slowly being removed. This is illustrated in Fig. 2.14.

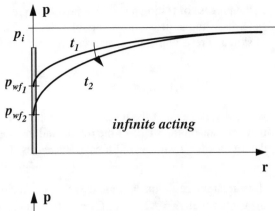

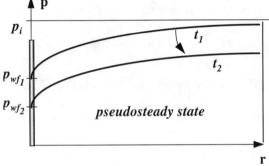

Figure 2.14

The condition of the reservoir during pseudosteady state is that pressure drop (everywhere) is due to the decompression of the reservoir fluid as fluid is produced from the well. This "volumetric" pressure loss is given simply from the definition of compressibility:

$$c_t = -\frac{1}{V}\frac{\Delta V}{\Delta p}$$

(2.34)

or, $\quad \Delta p = \frac{1}{V}\frac{q\Delta t}{c_t}$

(2.35)

where V is the total reservoir fluid volume, and the cumulative production ΔV is replaced by $q\,\Delta t$.

From Eq. 2.35 we can see that, during pseudosteady state, the pressure drop is:

(a) directly proportional to time --hence is identifiable as a straight line or a Δp vs. t or $\log(\Delta p)$ vs. $\log(\Delta t)$,

(b) dependent on reservoir volume -- hence is extremely useful as a means of estimating reservoir size.

In the process of testing a well, we provide an input **impulse** (usually a change in In terms of the dimensionless variables defined earlier, and incorporating the effects of skin and reservoir shape, the pressure drop can be written:

$$p_D = 2\pi t_{DA} + \tfrac{1}{2}\ln\left(\frac{2.2458A}{C_A r_w^2}\right) + s$$

(2.36)

where A is the area of the reservoir (or drainage region, and C_A is a shape factor that depends on the shape of the region and the position of the well. C_A has been determined for a variety of drainage shapes, as shown in Earlougher (1977).

If we treat the reservoir or drainage region as a circle of radius r_e, with the well at the center, then $A = \pi r_e^2$ and $C_A = 31.62$. In this case:

$$p_D = 2\pi t_{DA} + \tfrac{1}{2}\ln\left(0.472\frac{r_e}{r_w}\right) + s$$

(2.37)

Whatever the boundary shape, either Eq. 2.36 or 2.37 allow us to estimate the drainage area by determining the slope of the appropriate straight line found on a plot of Δp versus t. Substituting the definitions of p_D and t_{DA} in Eq. 2.36:

$$\Delta p = \frac{0.2342 qB}{(\phi c_t h)A}t + 70.65\frac{qB\mu}{kh}\left[\ln\left(\frac{2.2458A}{C_A r_w^2}\right) + 2s\right]$$

(2.38)

Thus the slope of the line Δp (psi) versus t (hrs) will be:

$$m_{Cartesian} = \frac{0.2342 qB}{(\phi c_t h)A}$$

(2.39)

Notice the groupings $\phi\, c_t\, h$, which is known as the "storativity," and $\phi\, h\, A$, which is the total pore volume within the drainage region.

Before seeking a straight line on a Cartesian pressure versus time graph, it is important to note that the pseudosteady state response does not appear until a certain value of t_{DA}. Recognition of this fact may be helpful to prevent the analysis of an improper Cartesian straight line. The latest time t_{DA} at which we can use the infinite system solution with less than 1% error can be considered the end of the infinite acting behavior (the end of the semilog straight line); this may be sooner than the start of pseudosteady state (the start of the Cartesian straight line). For the circular reservoir with the well at the center, the pseudosteady state starts *exactly* at the time the infinite acting behavior ends ($t_{DA} = 0.1$), hence the two times appear to overlap slightly (because of the 1% tolerance). However for more elongated shapes, or eccentric well locations, the infinite acting response ends long before pseudosteady state begins. The intervening period is a **transition period**.

Sometimes this transition period can be very long (for example the 5-to-1 rectangle, where it is sixteen times longer than the entire infinite acting response). Hence it would be a mistake to assume that the pseudosteady state response (Cartesian straight line) occurs immediately at the end of the infinite acting response (semilog straight line).

2.8.2 Fault Boundaries

Fault boundaries usually act as impermeable barriers, and therefore the pressure response of a well close to a single linear fault can begin to look like the response of a closed reservoir. However, the response is actually different. Since the well responds to only one boundary instead of being completely closed in on all sides, there is no pseudosteady state (at least not initially). Due to the influence calculable by superposition (described later), the well "sees itself in the mirror", and the net late time response is that of two identical wells. The semilog straight line of the original infinite acting response will therefore undergo a doubling in slope at the time the boundary effect is felt. This will be described later in "2.13 Superposition" on page 46.

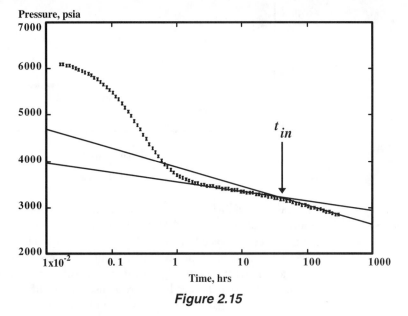

Figure 2.15

The distance d from the well to the fault can be estimated based on the time t_{in} at which the semilog straight line intersects the straight line of double slope. This point is illustrated in Fig. 2.15. Earlougher and Kazemi (1980) derived the following equation for the estimation of d:

$$d = 0.0122 \sqrt{\left(\frac{kt_{in}}{\phi \mu c_t} \right)}$$

(2.40)

2.8.3 Constant Pressure Boundaries

When the reservoir pressure is supported by fluid encroachment (either due to natural influx from an aquifer or gas cap, or by fluid injection) then a constant pressure boundary may be present. Such a boundary may completely enclose the well (as, for example, for a production well surrounded by injectors) or may be an open boundary to one side of the well (for example, in the case of an isolated producer/injector well pair). The effect of *any* constant pressure boundary will ultimately cause the well pressure response to achieve **steady state**, at which the well pressure will be the same constant pressure as the boundary, as in Fig. 2.16.

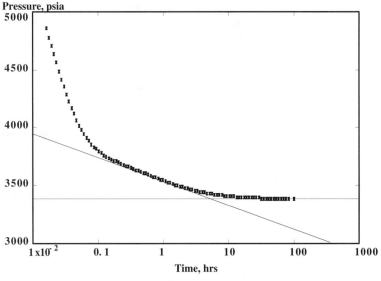

Figure 2.16

For the case of a ***circular*** constant pressure boundary with the well at the center, the wellbore pressure response will depart from the semilog straight line (infinite acting radial flow) response at a time t_{DA} of 0.08, pass through a transition period, and achieve steady state at a time t_{DA} of 0.3. Notice that this is rather later than the time required to achieve pseudosteady state in the case of impermeable boundaries. The time required to reach the steady state response will be different for drainage regions of different shape, and steady state will generally be reached sooner if some part of the constant pressure boundary is closer to the well.

For a ***linear*** constant pressure boundary, the distance d from the well to the fault can be estimated based on the time t_{in} at which the semilog straight line intersects the straight line of zero slope. This calculation is the same as that for a linear fault boundary, as shown earlier in Eq. 2.40.

2.9 Radius of Investigation

Having discussed reservoir boundary effects, it is worthwhile to define the concept of radius of investigation. The pressure response supposedly follows a diffusion type of response (Eq. 2.1), which would imply that a pressure change at the well would be felt at least infinitesimally everywhere in the reservoir. However, from a practical standpoint there will be some point distant from the well at which the pressure response is so small as to be undetectable. The closest such point defines the region of the reservoir that has been "tested" during the well test, and we may refer to the distance to this point as the **radius of investigation**. Since the definition of the radius of investigation depends on the definition of an "undetectable" pressure response, there have been a variety of definitions in the literature. Rather than make a comprehensive list or show preference to one of the definitions, it is better to acknowledge that radius of investigation is at best a vague concept that should only be used in a qualitative manner.

One way to think about the radius of investigation is in terms of the time to pseudosteady state. For example, it was discussed in "2.8.1 Closed Boundaries" on page 30 that in a circular reservoir the boundary effect (pseudosteady state) will be reached in a dimensionless time t_{DA} of 0.1. If no boundary effect has been seen at a particular time during a test, then the dimensionless time t_{DA} for the radius of investigation at that moment must be less than 0.1. This is the basis of the definition described by van Poollen (1964). Another way to define the radius of investigation was described in the book by Lee (1982). The basis of this approach is to consider the Green's function response (instantaneous point source response) and to determine the radius at which the pressure disturbance is maximized at any given time.

Both approaches give rise to similar expressions for the radius of investigation, which we might write here as:

$$r_{inv} = 0.03 \sqrt{\frac{kt}{\phi \mu c_t}}$$

$$(2.41)$$

Due to the nature of the concept, the coefficient 0.03 is likely to be seen with various values in different definitions. This serves to underline the fact that radius of investigation is not a parameter we may know with any degree of certainty.

As an example of the use of radius of investigation, consider the example data from Drawdown Example 1 shown earlier in "2.6.1 Semilog Analysis -- Illustrative Example" on page 25. The test duration t was 21.6 hours long, so the radius of investigation reached during the test would be about:

$$r_{inv} = 0.03 \sqrt{\frac{(77.1)(21.6)}{(0.21)(0.92)(8.72 \times 10^{-6})}} = 943 \ feet$$

2.10 Fractured Wells

In the USA, the ***majority*** of new oil and gas wells are hydraulically fractured as a routine part of their completion. In such a hydraulic fracturing procedure, the usual objective is the production of a single vertical fracture that completely penetrates the thickness of the productive formation and which extends some distance from the well. In other kinds of reservoir (for example geothermal wells which are commonly drilled in volcanic rocks) such major fractures may occur naturally and may be intersected by the well during drilling.

The fracture has much greater permeability than the formation it penetrates, hence it influences the pressure response of a well test significantly. Due to the ***linear*** geometry of the fracture, we will observe a pressure response other than the infinite-acting ***radial*** flow behavior discussed in Section 2.4, at least for part of the response. The interpretation of well tests from such wells must therefore consider the effects of the fracture; indeed, often tests of fractured wells are conducted specifically to determine fracture properties so that the effectiveness of the fracture stimulation operation can be evaluated.

The case of common practical interest is of a vertical fracture of length x_f, fully penetrating the formation (Figure 2.17).

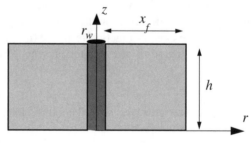

Figure 2.17

For the purposes of fractured well analysis, we often make use of a dimensionless time t_{Dxf} based on the fracture length x_f:

$$t_{Dxf} = \frac{0.000264kt}{\phi \mu c_t x_f^2}$$

$$(2.42)$$

Notice that this is directly related to the usual dimensionless time t_D as follows:

$$t_{Dxf} = t_D \frac{r_w^2}{x_f^2}$$

$$(2.43)$$

In well test analysis, three main fracture types are commonly considered: (a) finite conductivity fractures, (b) infinite conductivity fractures, and (c) uniform flux fractures.

2.10.1 Finite Conductivity Fractures

The most general case of a finite conductivity fracture was considered by Cinco, Samaniego and Dominguez (1978) and Cinco and Samaniego (1981a). Due to the linear flow in the fracture, different flow regimes can be observed at different times (Figure 2.18).

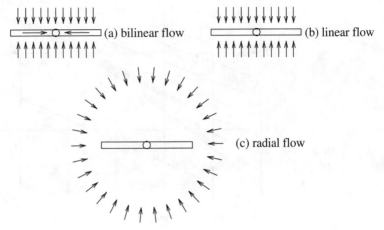

Figure 2.18

At early time, there is linear flow within the fracture and linear flow into the fracture from the formation (Figure 2.18 a). The combination of these two linear flows gives rise to the **bilinear** flow period. This part of the response is characterized by a straight line response with slope ¼ at early time on a log-log plot of pressure drop against time (Figure 2.19) since the pressure drop during this period is given by:

$$p_D = \frac{2.451}{\sqrt{k_{fD}w_{fD}}}t_{Dxf}^{1/4}$$

(2.44)

where the dimensionless fracture permeability and width are given respectively by:

$$k_{fD} = \frac{k_f}{k}$$

(2.45)

$$w_{fD} = \frac{w}{x_f}$$

(2.46)

where w is the width (or aperture) of the fracture.

Following the bilinear flow period, there is a tendency towards **linear** flow (Figure 2.18 b), recognizable by the upward bending in Fig. 2.19 towards a ½ slope on the log-log plot. In practice, the ½ slope is rarely seen except in fractures where the conductivity is infinite. Finite conductivity fracture responses generally enter a transition after bilinear flow (¼ slope), but reach **radial** flow (Fig. 2.18 c) before ever achieving a ½ slope (linear flow). Fig. 2.20 shows an example of such a response.

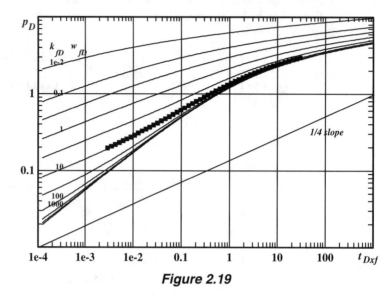

Figure 2.19

2.10.2 Infinite Conductivity Fractures

If the product $k_{fD}w_{fD}$ (defined by Eqs. 2.45 and 2.46) is larger than 300, then the fracture conductivity can be considered to be infinite. Such highly conductive fractures are quite possible in practice, especially in formations with lower permeability. The pressure response of a well intersecting an infinite conductivity fracture is very similar to that of the more general finite conductivity fracture case, except that the bilinear flow period is not present. An infinite conductivity fracture response is characterized by a truly linear flow response (Fig. 2.17 b), during which the pressure drop is given by:

$$p_D = \left(\pi t_{Dxf}\right)^{\frac{1}{2}}$$

(2.47)

Such a response shows as a ½ slope straight line on a log-log plot of pressure drop against time (Fig. 2.21), as shown by Gringarten, Ramey and Raghavan (1974).

Beyond the linear flow period, the response will pass through a transition to infinite acting radial flow (semilog straight line behavior).

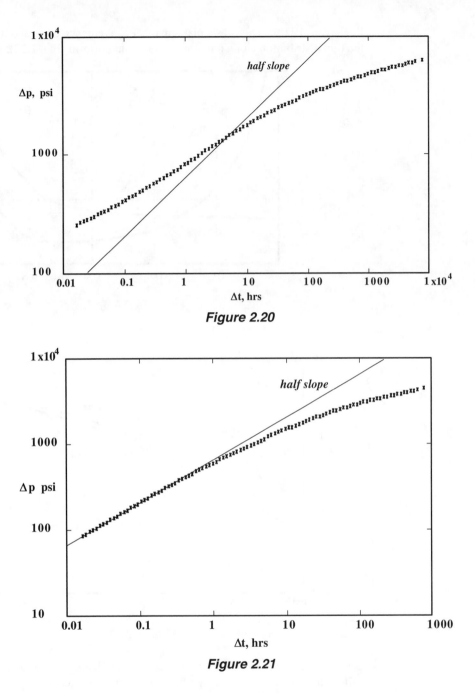

Figure 2.20

Figure 2.21

2.10.3 Uniform Flux Fracture

One of the earliest mathematical solutions to a fractured well problem assumed that the flow into the fracture was uniform along its length, Gringarten, Ramey and

---

Raghavan (1972). This was a mathematical convenience only, and it is known that the flux distribution along a fracture is far from uniform (Fig. 2.22).

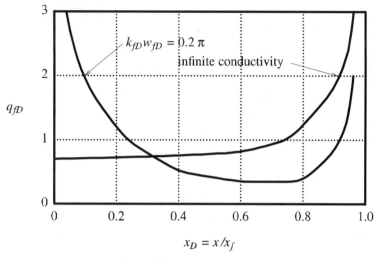

Figure 2.22

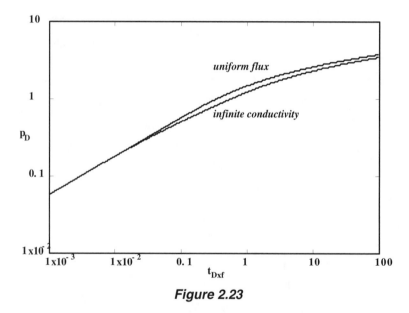

Figure 2.23

The only practical case in which the fracture flux distribution approaches uniformity is when there is a significant fracture "skin", caused by a lower permeability region between the fracture and the reservoir, Cinco and Samaniego (1981b). Whether this "damaged fracture" case is common or not is unclear, however, it is true that some wells appear to fit the uniform flux fracture response better than the infinite conductivity fracture response. In fact, there is only a rather small difference between the two flow situations (Fig. 2.23).

2.11 Dual Porosity Behavior

Discussion in previous sections has been focused on reservoirs with homogeneous properties. Because of the diffusive nature of pressure transmission, many reservoirs do indeed behave as if they were homogeneous, even though it is certain that the reservoir properties must be non-uniform to some extent. However, there is a type of reservoir heterogeneity that is noticeable in pressure transients in reservoirs that have distinct primary and secondary porosity. These pressure effects are known as **dual porosity** or **double porosity** behavior, and are quite commonly seen, particularly in naturally fractured reservoirs.

In a dual porosity reservoir, a porous "matrix" of lower transmissivity (primary porosity) is adjacent to higher transmissivity medium (secondary porosity), as in Figure 2.24.

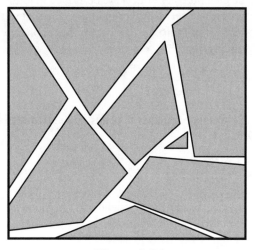

Figure 2.24

For the most part, we will discuss dual porosity reservoirs in terms of fractures and porous matrix blocks, however it is important to recognize that thin stratigraphic sequences of differing permeability can also give rise to dual porosity effects. Although there are variations, a common model is to associate *fluid storativity* with both the fractures (secondary porosity) and the matrix (primary porosity), but to assume *transmissivity* mainly in the fractures. In such a dual porosity model, fluid flows to the wellbore through the fractures alone, although may feed from the matrix blocks into the fractures.

Since there are now two interconnected media, we must define properties for each. The matrix permeability, porosity and total system compressibility are k_m, ϕ_m and c_{tm} respectively, while those for the fracture are k_f, ϕ_f and c_{tf}. In fractured systems, the fracture porosity ϕ_f may be very low, since the fracture volume is usually only a very small part of the total. Fracture compressibility c_{tf} is often very large due to the inflation/deflation effect as pressure changes in the fracture.

Based on the mathematical development (not shown here) of the equations of flow in the two media the normal dimensionless pressure and time are modified. Dimensionless pressure is based on fracture transmissivity, while dimensionless time is based on total (fracture plus matrix) storativity:

$$p_D = \frac{k_f h}{141.2qB\mu} \Delta p$$

(2.48)

$$t_D = \frac{0.000264k_f t}{\left(\phi_f c_{tf} + \phi_m c_{tm}\right)\mu r_w^2}$$

(2.49)

The dual porosity effects are described in terms of two parameters that relate primary and secondary properties. The first of the two parameters is the **storativity ratio**, ω, that relates the secondary (or fracture) storativity to that of the entire system:

$$\omega = \frac{\phi_f c_{tf}}{\phi_f c_{tf} + \phi_m c_{tm}}$$

(2.50)

The second parameter depends on the **transmissivity ratio**, and is designated as λ:

$$\lambda = \alpha \frac{k_m}{k_f} r_w^2$$

(2.51)

Here α is a factor that depends on the geometry of the interporosity flow between the matrix and the fractures:

$$\alpha = \frac{A}{xV}$$

(2.52)

where A is the surface area of the matrix block, V is the matrix volume, and x is a characteristic length. If the matrix blocks are cubes or spheres, then the interporosity flow is three-dimensional and λ is given by:

$$\lambda = \frac{60}{x_m^2} \frac{k_m}{k_f} r_w^2$$

(2.53)

where x_m is the length of a side of the cubic block, or the diameter of the spherical block. If the matrix blocks are long cylinders, then the interporosity flow is two-dimensional and λ is given by:

$$\lambda = \frac{32}{x_m^2} \frac{k_m}{k_f} r_w^2$$

(2.54)

where x_m is now the diameter of the cylindrical block. If the matrix blocks are slabs overlying each other with fractures in between (this kind of dual porosity often occurs in layered formations), then the interporosity flow is one-dimensional, and λ is given by:

$$\lambda = \frac{12}{h_f^2} \frac{k_m}{k_f} r_w^2$$

(2.55)

where h_f is the height of the secondary porosity slab (i.e., the thickness of the high permeability layer).

Values of ω can be less than or equal to one. The special case of $\omega = 1$ occurs when the matrix porosity is zero, hence refers to a reservoir that is **single porosity** -- the standard single porosity reservoir behaviors already discussed are simply special cases of dual porosity behavior with $\omega = 1$. In naturally fractured reservoirs, ϕ_f is usually very small, however, the large fracture compressibility c_{tf} means that ω is commonly less than 0.1 but not necessarily many orders of magnitude less. Values of λ are usually very small (for example, 10^{-3} to 10^{-10}). If the value of λ is larger than 10^{-3}, the level of heterogeneity is insufficient for dual porosity effects to be of importance, and again the reservoir acts as a single porosity.

Due to the two separate "porosities" in the reservoir, the dual porosity system has a response that may show characteristics of both of them. The secondary porosity (fractures), having the greater transmissivity and being connected to the wellbore, responds first. The primary porosity does not flow directly into the wellbore and is of lower transmissivity, therefore responds much later. The combined effect of the two gives rise to two separate semilog straight line responses, Figure 2.25.

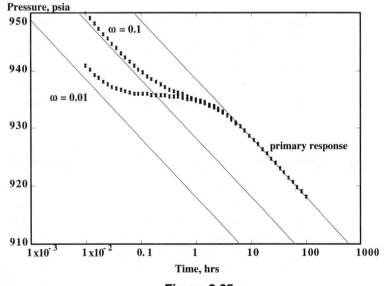

Figure 2.25

The separation between the straight lines is dependent on ω. For each log cycle that separates the two lines, ω is reduced by a factor of 10. Thus in Fig. 2.25 the lines for $\omega = 0.1$ are separated by one log cycle in time, and the lines for $\omega = 0.01$ are separated by two log cycles in time. The time at which the transition between them occurs is dependent on λ, Fig. 2.26. The value of λ can be estimated by locating the transition line, and then examining the *start* of the primary straight line (which is the later one) given by $t_D = (1 - \omega)/7\lambda$.

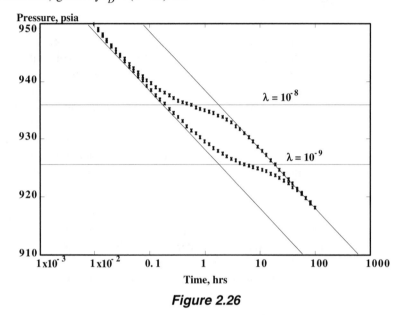

Figure 2.26

It should be stressed that this double straight line behavior is easily disguised by effects of wellbore storage (which may hide the secondary porosity transient completely) or by boundary effects (which may have an effect on the later transient before the primary porosity behavior is evident). In a modern analysis it is much more common to estimate ω and λ using the pressure derivative plot, as described in "3.3.3 Estimating Dual Porosity Parameters on Derivative Plots" on page 82.

2.12 Summary of Responses in Time Sequence

As has been discussed in the preceding five sections, a well test response may have different behavior at different times. The earliest time response is usually wellbore storage. Somewhat later, the response of fractures or primary porosity may be evident. At intermediate times, infinite acting radial flow may appear. Finally, late time responses may show the effects of reservoir boundaries. This combination of responses may give rise to an overall transient such as in Figure 2.27. Clearly there are many possible combinations of effects, as can be imagined by examination of Table 2.1.

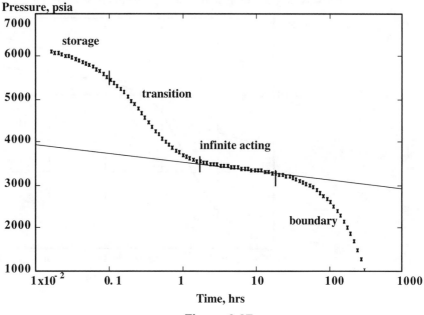

Figure 2.27

	Early time	**Intermediate time**	**Late time**
radial flow	storage	infinite-acting radial flow	closed boundary sealing fault constant pressure
fractures	storage bilinear flow	radial flow	closed boundary sealing fault constant pressure
dual porosity	storage	dual porosity behavior transition radial flow	closed boundary sealing fault constant pressure

Table 2.1

Recognition and appropriate analysis of these different responses is the key to proper interpretation of a well test. Data from a real well test can start and end at any time, so one or more of the responses can be missing. Also, depending on parameter values, one response may overlap and hide another. For example, in a dual porosity reservoir, the first semilog straight line characteristic of the secondary porosity may be completely hidden by wellbore storage (Figure 2.28).

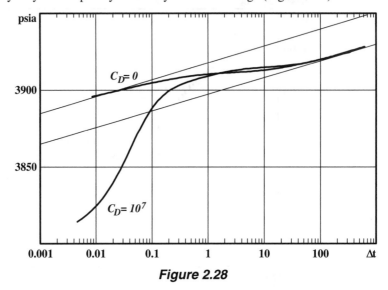

Figure 2.28

2.13 Superposition

One of the most powerful techniques in reservoir engineering is **superposition**. This approach makes it possible to construct reservoir response functions in complex situations, using only simple basic models. Superposition is especially useful in well test analysis, since we can use it to represent the response due to several wells by adding up the individual well responses. By appropriate choice of flow rate and well location, we can also represent various reservoir boundaries. In addition, we can use superposition in time to determine the reservoir response to a well flowing at variable rate, by using only constant rate solutions.

The principle of superposition is very simple. It says that the response of the system to a number of perturbations is exactly equal to the sum of the responses to each of the perturbations as if they were present by themselves. It should be noted in passing that the principle of superposition only holds for linear systems (in the mathematical sense), however these include most of the standard response functions used in well test analysis, such as the constant rate radial flow, dual porosity, fractured and bounded well solutions described earlier.

To begin to understand use of superposition, consider the pressure drop in the reservoir at point A due to the production of two wells at B and C (Figure 2.29).

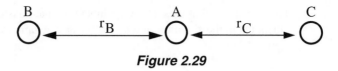

Figure 2.29

Assuming that wells B and C are both line source wells without wellbore storage, the pressure drop at A due to the production of both wells is (assuming $t_D > 10$):

$$\Delta p = \frac{141.2 q_B B \mu}{kh}\left[\frac{1}{2}\left(\ln\frac{0.000264kt}{\phi\mu c_t r_B^2}+0.80907\right)\right]$$

$$+\frac{141.2 q_C B \mu}{kh}\left[\frac{1}{2}\left(\ln\frac{0.000264kt}{\phi\mu c_t r_C^2}+0.80907\right)\right]$$

$$(2.56)$$

It can be confirmed from Eq. 2.56 that the total pressure drop is equal to the sum of the individual pressure drops. This is true for any number of wells.

Another interesting observation can be made if both wells produce at identical rate, and the point A is exactly midway between them. In this case the pressure *gradient* towards the other wells, thus the net flux towards either well, is zero. Hence *any* point midway between the wells is a no flux point, and we can replace all such points by an impermeable barrier without affecting the flow distribution or the pressure field (Figure 2.30).

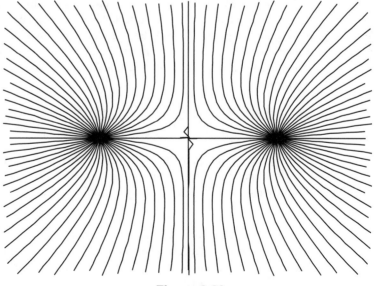

Figure 2.30

Alternatively, if the wells are equidistant, but have equivalent flow rates opposite in sign, then the pressure drop at the midpoint will be exactly zero, since the pressure drop due to one well will be exactly canceled by the pressure rise due to the other.

The net result is that all such midpoints remain at constant pressure, and the effect is identical to the situation in which a linear constant pressure boundary is present (Figure 2.31).

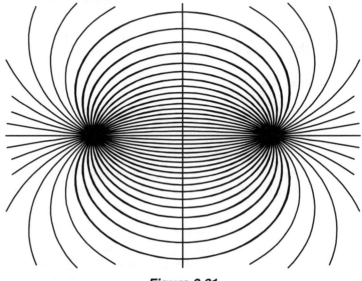

Figure 2.31

Thus we have discovered an important method of representing the effect of a boundary, using only very simple pressure drop solutions for wells *in an infinite reservoir*. The effect of an impermeable boundary can be replicated exactly by placing an "image" well at a distance from the original well that is exactly twice the distance of the boundary from the original well. Also, we can see from Eq. 2.56 that at late time the effect of two identical wells, measured at the original well, will be given by Eq. 2.57.

$$\Delta p = \frac{141.2 q_B B \mu}{kh} \left[\frac{1}{2} \left(\ln \frac{0.000264kt}{\phi \mu c_t r_W^2} + \ln \frac{0.000264kt}{\phi \mu c_t (2r_e)^2} \right) + 0.80907 \right]$$

(2.57)

where r_e is the distance to the impermeable boundary. Hence we can understand why the slope of the semilog straight line doubles as the boundary effect becomes significant (as was described earlier in Section 2.8.2, and is shown in Figure 2.32).

Why is the semilog slope not doubled for all time? Since r_e^2 is so much larger than r_w^2, the second well effect does not become significant until time t becomes large (recall that the logarithmic approximation to the exponential integral function is only valid for $t_D/r_D^2 > 10$).

Using the image well concept, it is very simple to create the effect of quite complex boundary shapes, including mixtures of impermeable and constant pressure boundaries, even in three dimensions. Consider for example the infinite array of images that can be used to create the effect of a rectangular boundary (Fig. 2.33).

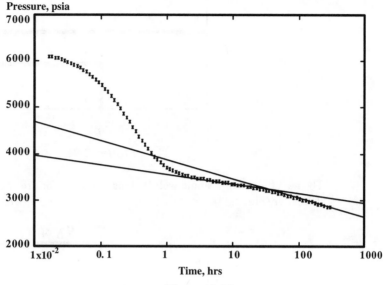

Figure 2.32

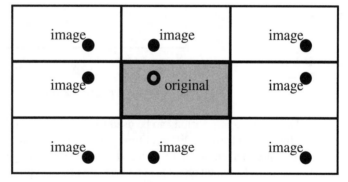

Figure 2.33

2.14 Time Superposition -- Multirate Tests

A second important use of superposition is to add together the effects of wells at different *times*. Consider for example the case of Fig. 2.28 where *A*, *B* and *C* are all at the same point. The pressure drop of two wells *at the same location*, one with flow rate q_B and the other with flow rate q_C, is identical to the pressure drop due to a single well with flow rate $q_B + q_C$. We can imagine replacing two small pumps by one big pump. However, let us go on to consider the case where the first well starts flowing at time zero, and the second well does not start flowing until time t_p (Fig. 2.34).

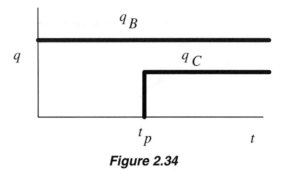

Figure 2.34

The net effect is of a single well flowing at rate q_B for time t_p, and at rate $q_B + q_C$ thereafter (Fig. 2.35).

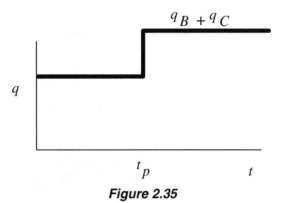

Figure 2.35

This same superposition can be used for any number of "wells", each with constant flow rates starting at different times, thus it is possible to generate the reservoir response to a single well flowing at variable rate, using only the same constant rate solutions described already. The variable rate is approximated by a series of "stairsteps," as in Fig. 2.36.

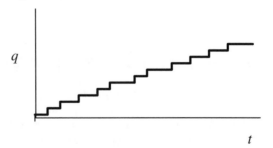

Figure 2.36

When a well test contains a series of different flow rates, or a continuously varying flow rate, the combination of the pressure transients due to the varying flow rate is called **convolution**. Convolution can be understood physically in terms of the principles of superposition just described, or mathematically in terms of Duhamel's Principle, which can be stated in the context of pressure transient computation as:

$$p_{wD} = \int_0^{t_D} q'_D(\tau)\left[p_D(t_D - \tau) + s\right]d\tau \tag{2.58}$$

or,

$$p_{wD} = \int_0^{t_D} q_D(\tau)p'_D(t_D - \tau)d\tau + sq_D(t_D) \tag{2.59}$$

Where q_D is the flow rate relative to some reference value q_{ref}. It is important to note that q_{ref} will be the flow rate used to define p_{wD}. The prime designates the derivative with respect to time τ. These equations originate from the paper by van Everdingen and Hurst (1949).

The complement of convolution is **deconvolution**, in which the pressure response for constant rate production can be computed from the (measured) pressure response due to the actual (multirate) flow history. The rigorous mathematical treatment of both convolution and deconvolution usually require the application of computer-based methods, so full discussion will be deferred until later (see "3.7 Desuperposition" on page 97).

There are several traditional methods of handling convolution approximately in cases where the well flows in infinite acting radial flow (semilog pressure behavior). The following description follows that of Earlougher (1977). For a series of N constant step changes in flow rate, such as those shown in Fig. 2.36, the integral in Eq. 2.58 can be rewritten:

$$p_{wf}(t) = p_i - \frac{141.2B\mu}{kh}\left\{q_1\left[p_D(t_D) + s\right] + \sum_{j=2}^N (q_j - q_{j-1})\left[p_D(t_D - t_{jD}) + s\right]\right\} \tag{2.60}$$

Where q_j is the flow step between t_{j-1} and t_j. In cases where the response can be approximated by semilog behavior (infinite acting radial flow), the pressure from Eq. 2.60 can be written:

$$p_{wf}(t) = p_i - \frac{162.6B\mu}{kh}\left\{\sum_{j=1}^{N-1} q_j \log\left(\frac{t - t_{j-1}}{t - t_j}\right) + \right.$$
$$\left. q_N\left[\log(t - t_{N-1}) + \log\frac{k}{\phi\mu c_t r_w^2} - 3.2275 + 0.8686s\right]\right\} \tag{2.61}$$

This equation is more convenient to use if rewritten as:

$$\frac{p_i - p_{wf}(t)}{q_N} = \frac{162.6B\mu}{kh}\left\{\sum_{j=1}^{N}\left[\frac{q_j - q_{j-1}}{q_N}\log(t - t_{j-1})\right] + \right.$$

$$\left. \log\frac{k}{\phi\mu c_t r_w^2} - 3.2275 + 0.8686s\right\} \tag{2.62}$$

To use this equation correctly, it is important to notice than the value of N depends on the time t at which the pressure is determined. The flow rate q_N is the flow rate at the time t in Eq. 2.62. The same consideration regarding N and q_N also applies to Eq. 2.61.

Based on Eq. 2.62, a plot of $\dfrac{p_i - p_{wf}(t)}{q_N}$ vs. $\displaystyle\sum_{j=1}^{N}\left[\frac{q_j - q_{j-1}}{q_N}\log(t - t_{j-1})\right]$ should

show a straight line with slope $162.6B\mu/kh$ during infinite acting radial flow behavior. Such a plot is sometimes known as a **rate-normalized** plot or a **multirate superposition** plot.

As an example of the use of rate-normalization, Fig. 2.38 shows a normal semilog plot of the drawdown test with variable flow rate shown in Fig. 2.37. The rate-normalized plot in Fig. 2.39 shows the expected straight line behavior, whereas the semilog plot in Fig. 2.38 shows a continuously changing curve.

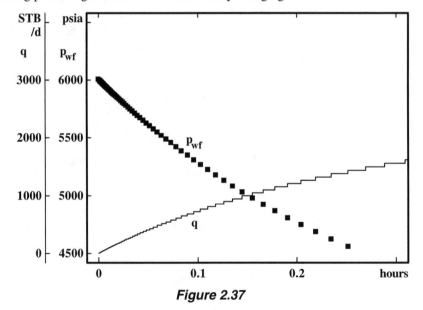

Figure 2.37

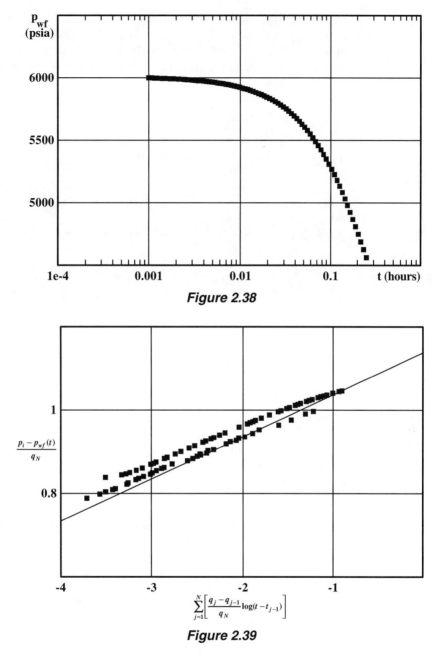

Figure 2.38

Figure 2.39

A similar but more accurate form of the rate-normalization idea was described by Meunier, Wittmann, and Stewart (1985), who improved the procedure by approximating continuously varying flow rate data as a series of linear "ramp" segments instead of as the series of constant steps used in Eq. 2.60 and depicted in Fig. 2.36. This **sandface rate convolution** method as described by Meunier, Wittmann, and Stewart (1985) still depends on the assumption that the reservoir is responding with semilog behavior (infinite acting radial flow). In the sandface rate convolution method, Eq. 2.58 is approximated by replacing the flow rate function by a series of linear "ramps" between adjacent flow rate measurements, and the pressure function is approximated using the semilog function described by Eq. 2.27 or 2.28, such that:

$$\frac{p_i - p_{wf}(t_N)}{q_{DN}} = \frac{162.6 q_{ref} B\mu}{kh} \frac{X_N - q_{DN} \log e}{q_{DN}}$$

$$+ \frac{162.6 q_{ref} B\mu}{kh} \left[\log \frac{k}{\phi\mu c_t r_w^2} - 3.2275 + 0.8686 s \right]$$

$$(2.63)$$

where:

$$X_N = \sum_{j=1}^{N-1} \left(\frac{q_{Dj+1} - q_{Dj}}{t_{j+1} - t_j} - \frac{q_{Dj} - q_{Dj-1}}{t_j - t_{j-1}} \right) (t_N - t_j) \log(t_N - t_j) +$$

$$\frac{q_{D1} - q_{D0}}{t_1 - t_0} t_N \log t_N$$

$$(2.64)$$

$$q_{D_N} = q(t_N) / (q_{ref} B)$$

$$(2.65)$$

It should be noted that this procedure will fail if any of the individual flow rates are zero.

2.15 Buildup Tests

A particular case of time superposition that is of practical importance is where there are only two flow rates, such as when q_B is q, starting at time zero, and q_C is $-q$, starting at the time t_p.

The effect of these two flow rates is the representation of a well which is produced for a time t_p, at rate q, and then shut in (Figure 2.40).

This gives us a means to generate the pressure response during a buildup test, using the simple constant rate solutions generated for drawdown tests (as in Eq. 2.66).

$$p_D(t_D) = p_D(t_{pD} + \Delta t_D) - p_D(\Delta t_D)$$

$$(2.66)$$

This is illustrated in Fig. 2.41, and is true regardless of the reservoir model used.

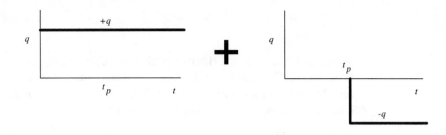

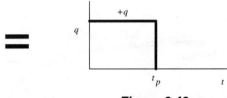

Figure 2.40

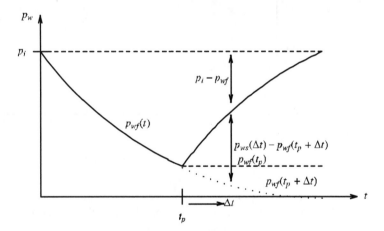

Figure 2.41

While discussing this point, we can observe that this time superposition leads to a particularly simple result during infinite acting radial flow. During this flow regime:

$$P_{ws} = p_i - \frac{141.2qB\mu}{kh}\left[\frac{1}{2}\left(\ln\frac{k(t_p + \Delta t)}{\phi\mu c_t r_W^2} - \ln\frac{k\Delta t}{\phi\mu c_t r_W^2}\right)\right]$$

(2.67)

or, $$P_{ws} = p_i - \frac{162.6qB\mu}{kh}\log\frac{(t_p + \Delta t)}{\Delta t}$$

(2.68)

Thus a plot of pressure against the logarithm of $(t_p+\Delta t)/\Delta t$ will show a straight line of slope:

$$m = \frac{162.6qB\mu}{kh}$$

(2.69)

Such a plot is known as a **Horner plot**, and we refer to $(t_p+\Delta t)/\Delta t$ as the **Horner time** (Fig. 2.42). Due to the definition of Horner time, it should be noted that actual time increases to the left in Fig. 2.42. As the shut-in time Δt tends to infinity, the Horner time $(t_p+\Delta t)/\Delta t$ tends to 1.

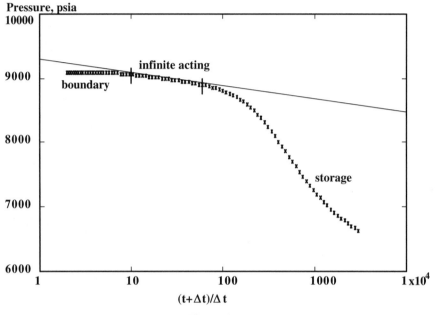

Figure 2.42

The Horner plot may also be used to estimate the skin factor. Since the skin factor is a dimensionless pressure drop, the skin effect only influences the flowing period of the test. Thus it is necessary to include the data point representing the last flowing pressure -- this point is $p_{wf}(t_p)$. The difference between the shut-in pressure and the last flowing pressure, *assuming infinite-acting radial flow*, is:

$$p_{ws}(t_p + \Delta t) - p_{wf}(t_p) = \tfrac{1}{2} \frac{141.2qB\mu}{kh}$$

$$\left[\ln \frac{t_p \Delta t}{t_p + \Delta t} + 0.80907 + \ln \frac{0.000264k}{\phi \mu c_t r_w^2} \right] + s$$

(2.70)

If we substitute a value of $\Delta t = 1\ hour$, then we can obtain an estimate of the skin factor:

$$s = 1.151 \left[\frac{p_{1hr} - p_{wf}}{m} - \log \frac{kt_p}{(t_p + 1)\phi \mu c_t r_w^2} + 3.2274 \right]$$

(2.71)

As with drawdown tests, it is important to note that the value of p_{1hr} needs to be taken *from the Horner straight line*, or the extrapolation of it This is due to the assumption of infinite-acting radial flow -- the flow regime at the arbitrary time of 1 hour may not be infinite-acting, and therefore the actual pressure data point at this time is not the correct one to use.

The Horner time is useful for semilog analysis, however we still have problems with log-log (type curve) analysis. This is because in type curve analysis of a drawdown test, we plot log $(p_i - p_{wf})$ against log Δt. We could do this correctly for shut-in tests if we could plot the logarithm of $p_{ws}(\Delta t)$ - $p_{wf}(t_p + \Delta t)$ (see Fig. 2.41), by subtracting the "would have been" flowing pressure from the measured well pressure. Unfortunately we do not know what this "would have been" pressure is. Agarwal (1980) developed the concept of an "equivalent time" (which is known as the **Agarwal equivalent time**) as a time at which the measurable pressure difference $p_{ws}(\Delta t)$ - $p_{wf}(t_p)$ is equal to the correct pressure difference $p_{ws}(\Delta t)$ - $p_{wf}(t_p + \Delta t)$. This equivalent time can be determined exactly for infinite acting radial flow, when the log approximation is valid (i.e., when $t_D > 10$). For the time to be equivalent, we require that:

$$p_{ws}(t_{equiv}) - p_{wf}(t_p) = p_{ws}(\Delta t) - p_{wf}(t_p + \Delta t) \tag{2.72}$$

As long as each pressure varies as log t, then we can determine that:

$$t_{equiv} = \frac{t_p \Delta t}{t_p + \Delta t} \tag{2.73}$$

This is the Agarwal equivalent time, and using it in place of Δt will allow drawdown type curves to be used for buildup. This is strictly true only for infinite acting radial flow without wellbore storage, however has been shown to work in some cases even in tests which contain storage or fracture effects.

The Agarwal effective time concept can be generalized to take account of any number of flow periods preceding the shutting of the well. This generalization follows a derivation similar to that shown in Eqs. 2.58-2.62. If we have measured a series of N different flow rates prior to shut-in, the well shut-in pressure *assuming infinite acting (semilog) behavior* can be written:

$$p_{ws} = p_i - \frac{141.2 q_N B\mu}{kh}\left[\frac{1}{2}\sum_{j=1}^{N}\frac{q_j - q_{j-1}}{q_N}\left(\ln\frac{k(t_j - t_{j-1} + \Delta t)}{\phi\mu c_t r_W^2} - \ln\frac{k\Delta t}{\phi\mu c_t r_W^2}\right)\right] \tag{2.74}$$

Notice here that q_N is the last rate the well flowed at before being shut. Thus instead of plotting a semilog plot against Horner time, we may plot pressure against the log time variable:

$$\sum_{j=1}^{N} \frac{q_j - q_{j-1}}{q_N} \log \left[\frac{\left(t_j - t_{j-1} + \Delta t\right)}{\Delta t} \right]$$

(2.75)

Such a plot would then have the familiar slope m, given by Eq. (2.69).

A discussion of the use and limitations of these types of semilog superposition approaches may be found in Larsen (1983).

2.15.1 Buildup Tests -- Illustrative Example

Consider the example buildup data shown in Fig. 2.43. The well has been shut in after 21.6 hours of production. The data have been plotted Δp vs. Δt on log-log coordinates as the first step we would usually perform for semilog analysis.

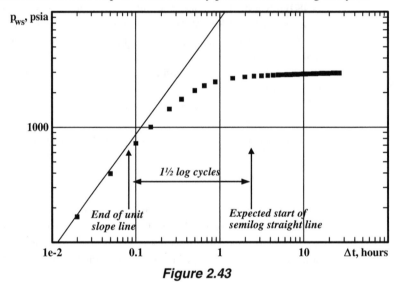

Figure 2.43

The reservoir behavior is expected to start after a shut-in time Δt of about 2 hours. This would correspond to a Horner time of $(t_p + \Delta t)/\Delta t = (21.6 + 2)/2 = 11.8$. Plotting a Horner plot as in Fig. 2.44, the Horner straight line is seen to start at around this point. The slope of the Horner straight line is 249.1 psi/log cycle.

Other parameters for this test are as follows:

B_t (RB/STB)	1.21	r_w (feet)	0.401
μ_o (cp)	0.92	h (feet)	23
c_t (/psi)	8.72×10^{-6}	p_{wf} (psia)	2989.39
ϕ	0.21	t_p (hours)	21.6
q (STB/d)	2500		

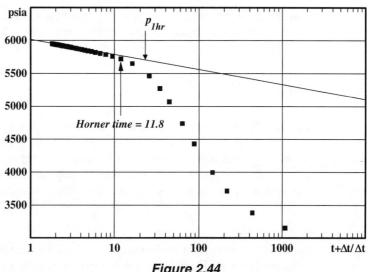

Figure 2.44

Based on the slope of the straight line, given by Eq. 2.69, and the values of the other well and reservoir parameters (which are mainly the same as in Drawdown Example 1 shown in "2.6.1 Semilog Analysis -- Illustrative Example" on page 25), we can estimate the permeability:

$$m = \frac{162.6qB\mu}{kh}$$

$$249.1 = \frac{162.6(2500)(1.21)(0.92)}{k(23)}$$

Hence the permeability $k = 79.0$ md.

The skin factor can be estimated based on the p_{1hr} point on the Horner straight line. A shut-in time Δt of 1 hour corresponds to a Horner time of $(t_p+\Delta t)/\Delta t = (21.6 + 1)/1 = 22.6$. As with Drawdown Example 1, the p_{1hr} point lies on the Horner straight line extension at a pressure of 5677.47 psia and does not lie within the actual data (at a shut-in time of 1 hour the pressure was actually 5461.4 psia). Using Eq. 2.71:

$$s = 1.151\left[\frac{p_{1hr} - p_{wf}}{m} - \log\frac{kt_p}{(t_p + 1)\phi\mu c_t r_w^2} + 3.2274\right]$$

$$s = 1.151\left[\frac{5677.47 - 2989.39}{249.1} - \log\frac{(79.0)(21.6)}{(21.6 + 1)(0.21)(0.92)(8.72 \times 10^{-6})(0.41)^2} + 3.2274\right]$$

Hence the skin factor $s = 6.44$.

This simple buildup illustration does not encounter the more difficult situation in which boundary effects may be present during the buildup or during the preceding drawdown. When boundary effects are present the analysis can be considerably more problematic since the Horner straight line may be difficult to identify.

2.15.2 Treating Buildups as Drawdowns

As discussed earlier, it may be possible to treat a buildup the same as a drawdown if the preceding production time is three to five times as long as the period of shut in, although some care is required if boundary effects are present in either the drawdown or buildup responses. With computer-aided interpretation, it is not difficult to accommodate the full history of the production prior to the buildup, so the need or interest in using traditional drawdown analysis methods is much reduced.

A case of more practical interest is one in which the previous production history is unknown (perhaps due to collective production through a manifold, for example), and where the well has been in production for a long period of time. In such a case, any transient attributable to the production is likely to be very small during the duration of the buildup test. Thus the buildup test may be reasonably treated as if an injection of fluid at the rate $-q$ started at the time of the test, where q was the stabilized rate of production prior to shut in. Rather than include the production history back to the first flow of the well, it is adequate to treat the transient as if it had started just at the time of shut in.

2.15.3 Average Reservoir Pressure

One common objective for a shut-in test is to estimate the average reservoir pressure, which is expected to change as reservoir production occurs.

Finding the "initial" reservoir pressure p_i can be an integral part of nonlinear regression analysis, whether the test is drawdown or buildup type. In a buildup analysis, the estimated value of p_i obtained will be the estimate of average reservoir pressure. Even without nonlinear regression, the ability to "simulate" extended duration shut-in tests makes it possible, in computer-aided interpretation, to extrapolate to the new reservoir pressure p_i, if the drainage shape is known or assumed. However, the traditional methods will be described here for completeness.

The situation is straightforward if the actual initial pressure before production began is known, since the average reservoir pressure is then a simple matter of material balance, based on the equation for pseudosteady state behavior:

$$\frac{kh}{141.2qB\mu}(p_i - \overline{p}) = 2\pi t_{DA}$$

$$(2.76)$$

Notice that it is necessary to know the drainage area, and that no information from the actual well test is used (except the producing time).

If the initial p_i is unknown, which may be the case if the well has been in production for a long time, then a graphical technique due to Matthews, Brons and Hazebroek (1954) is used. An example of an MBH plot is shown in Fig. 2.45.

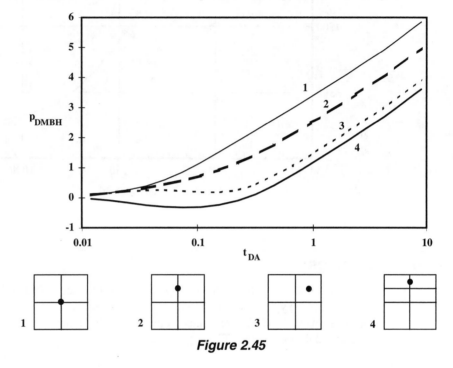

Figure 2.45

The procedure is to determine t_{DA}, using the value of producing time t_p, and an estimate or assumption of the drainage area. Next, the corresponding p_{DMBH} is read from the MBH plot (such as Fig. 2.45), making use of a prior estimate or assumption about the shape and/or configuration of the boundary. Finally, the value of average reservoir pressure, $\bar{p}$, is determined from the definition of p_{DMBH}:

$$p_{DMBH} = 2\frac{kh}{141.2\,qB\mu}(p^* - \bar{p})$$

(2.77)

$$p_{DMBH} = \frac{2.303(p^* - \bar{p})}{m}$$

(2.78)

where m is the slope of the Horner straight line, and p^* is the point at which the extension of the Horner straight line meets the axis $(t_p + \Delta t)/\Delta t = 1$, as in Fig. 2.46. This point is sometimes known as the **Horner extrapolated pressure** or the **Horner false pressure**.

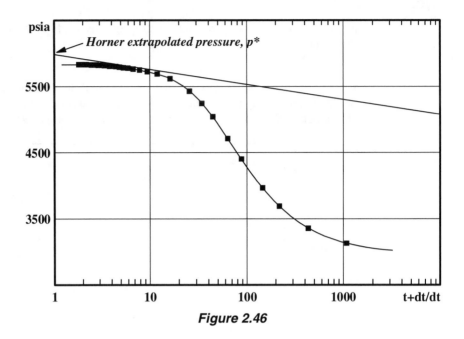

Figure 2.46

2.16 References

Agarwal, R.G.: "A New Method to Account for Producing Time Effects When Drawdown Type Curves are Used to Analyze Pressure Buildup and Other Test Data", paper SPE 9289 presented at the 55th SPE Annual Technical Conference and Exhibition, Dallas, TX, Sept. 21-24, (1980).

Cinco, H., Samaniego, F., and Dominguez, N.: "Transient Pressure Behavior for a Well with a Finite Conductivity Vertical Fracture", *Soc. Petr. Eng. J.*, (August 1978), 253-264.

Cinco, H., Samaniego, F.: "Transient Pressure Analysis for Fractured Wells", *J. Pet. Tech.*, (September 1981).

Cinco, H., Samaniego, F.: "Transient Pressure Analysis: Finite Conductivity Fracture Case Versus Damaged Fracture Case", paper SPE 10179 presented at the 6th SPE Annual Technical Conference and Exhibition, San Antonio, TX, Oct. 5-7, (1981b).

Earlougher, R.C., Jr.: *"Advances in Well Test Analysis"*, Society of Petroleum Engineers Monograph 5, Dallas, TX, (1977).

Earlougher, R.C., Jr., and Kazemi, H.: "Practicalities of Detecting Faults from Buildup Testing", *J. Pet. Tech.*, (Jan. 1980), 18-20.

Gringarten, A.C., Ramey, H.J., Jr., and Raghavan, R.: "Pressure Analysis for Fractured Wells", paper SPE 4051 presented at the 47th SPE Annual Technical Conference and Exhibition, San Antonio, TX, Oct. 8-11, (1972).

Gringarten, A.C., Ramey, H.J., Jr., and Raghavan, R.: "Unsteady State Pressure Distribution Created by a Well with a Single Infinite Conductivity Vertical Fracture", *Soc. Petr. Eng. J.*, (August 1974), 347-360.

Gringarten, A.C., and Ramey, H.J., Jr.: "An Approximate Infinite Conductivity Solution for a Partially Penetrating Line-Source Well", *Soc. Petr. Eng. J.*, (April 1975), 140-148; *Trans.* AIME, **259**.

Kuchuk, F.J., and Kirwan, P.A.: "New Skin and Wellbore Storage Type Curves for Partially Penetrating Wells", *SPE Formation Evaluation*, (Dec. 1987), 546-554.

Larsen, L.: "Limitations on the Use of Single- and Multiple-Rate Horner, Miller-Dyes-Hutchinson, and Matthews-Brons-Hazebroek Analysis", paper SPE 12135, Proceedings 1983 SPE Annual Technical Conference, San Francisco, Oct. 5-8, (1983).

Lee, W.J.: *"Well Testing"*, SPE Dallas, TX; SPE Textbook Series, No. 1, (1982).

Matthew, C.S., Brons, F., and Hazebroek, P.: "A Method for Determination of Average Reservoir Pressure in Bounded Reservoirs", *Trans.*, AIME, (1954), **201**, 182-191.

Meunier, D., Wittmann, M.J., and Stewart, G.: "Interpretation of Pressure Buildup Tests Using In-Situ Measurement of Afterflow", *J. Pet. Tech.* (Jan. 1985), 143-152.

van Everdingen, A.F., and Hurst, W.: "Application of the Laplace Transformation to Flow Problems in Reservoirs", *Trans.* AIME, 186 (1949), 305-324.

van Poollen, H.K.: "Radius-of-Drainage and Stabilization-Time Equations", *Oil and Gas J.*, (Sept. 14, 1964), 138-146.

3. COMPUTER-AIDED ANALYSIS

3.1 Overview

Well test analysis by traditional methods makes considerable use of graphical presentations. Much of the theory of the field has concentrated on procedures to aid graphical analysis, as has been explained in Section 2. It can be seen that many of the underlying principles are based upon the following restrictions:

(a) Drawdown in a single well

(b) Constant rate production

Using approximations that are often associated with log (time) behavior (i.e., infinite acting radial flow) there are some useful extensions that can be made to graphical techniques (for example to allow analysis of buildup and multirate tests).

The first objective in computer-aided analysis is to speed up the traditional graphical techniques by allowing rapid presentation of graphs and by performing the standard estimation calculations. However a more important objective is to extend the analysis beyond the restrictions inherent in traditional methods. Specifically, a computer-aided interpretation can accommodate situations that are either only approximated in traditional methods, or which cannot be handled at all, such as:

(a) Continuously varying rate

(b) Multiple wells

(c) Complex geometry

(d) Downhole measurements of flow rate

(e) Indefinite initial pressure

Thus a computer-aided interpretation allows the reservoir engineer to obtain better results in less time, to a wider range of field conditions. Economides, Joseph, Ambrose, and Norwood (1989), Gringarten (1986) and Horne (1994) have summarized modern approaches to well test analysis using computers.

The procedure in many computer-aided analyses is to follow traditional lines of approach as far as they are applicable, and then to extend the interpretation using

---

the additional capabilities permitted by the computer-aided approach. The advantage of starting the analysis along traditional lines is that the techniques are familiar to the engineer engaged in the work and the expertise gained over many years of traditional well interpretation is not abandoned. Thus computer-aided interpretation remains dependent on graphical presentations (in fact the simplest of computer assistance may be restricted to that).

3.2 Graphical Presentations

Data presented in a pictorial (graphical) form are much easier to understand than a simple table of numbers. A useful "toolbox" of graphing functions are therefore an essential part of a computer-aided well test interpretation system.

Flow Period	Characteristic	Plot Used
Infinite-acting radial flow (drawdown)	Semilog straight line	p vs. $\log \Delta t$, (semilog plot, sometimes called MDH plot)
Infinite-acting radial flow (buildup)	Horner straight line	p vs. $\log (t_p+\Delta t)/\Delta t$, (Horner plot)
Wellbore storage	Straight line p vs. t, or Unit slope $\log \Delta p$ vs. $\log \Delta t$	$\log \Delta p$ vs. $\log \Delta t$, (log-log plot, type curve)
Finite conductivity fracture	Straight line slope ¼, $\log \Delta p$ vs. $\log \Delta t$ plot	$\log \Delta p$ vs. $\log \Delta t$, or Δp vs. $\Delta t^{1/4}$
Infinite conductivity fracture	Straight line slope ½, $\log \Delta p$ vs. $\log \Delta t$ plot	$\log \Delta p$ vs. $\log \Delta t$, or Δp vs. $\Delta t^{1/2}$
Dual porosity behavior	S-shaped transition between parallel semilog straight lines	p vs. $\log \Delta t$, (semilog plot)
Closed boundary	Pseudosteady state, pressure linear with time	p vs. Δt (Cartesian plot[*])
Impermeable fault	Doubling of slope on semilog straight line	p vs. $\log \Delta t$, (semilog plot)
Constant pressure boundary	Constant pressure, flat line on all p,t plots	Any

* Although pseudosteady state response does show as a straight line on a Cartesian plot, it is very important that the Cartesian plot *not* be used to diagnose pseudosteady state. This is because almost all late time responses (including infinite-acting) appear to be straight on a Cartesian plot, which is deceptive for shallow slopes.

Table 3.1

The objective in examining a series of graphical presentations of the data is to correctly identify the different characteristic flow periods occurring during the well test. The various different flow periods described earlier in Section 2 are summarized in Table 3.1, along with their characteristic signature and the plot on which this signature is best recognized.

3.3 The Derivative Plot

From Table 3.1 it can be appreciated that different plots are used for different purposes, and most analysis will require the consideration of several plots.

Modern analysis has been greatly enhanced by the use of the **derivative plot** introduced by Bourdet, Whittle, Douglas and Pirard (1983), also discussed in Bourdet, Ayoub and Pirard (1989). The derivative plot provides a simultaneous presentation of log Δp vs. log Δt and log tdp/dt vs. log Δt, as in Figure 3.1.

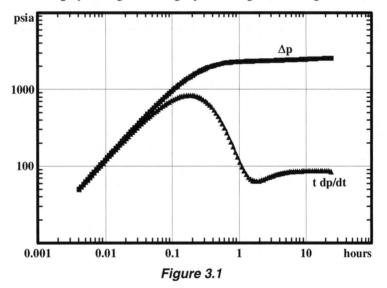

Figure 3.1

The advantage of the derivative plot is that it is able to display in a single graph many separate characteristics that would otherwise require different plots. These characteristics are shown in Figures 3.2 - 3.9, compared to their traditional plots (as listed in Table 3.1).

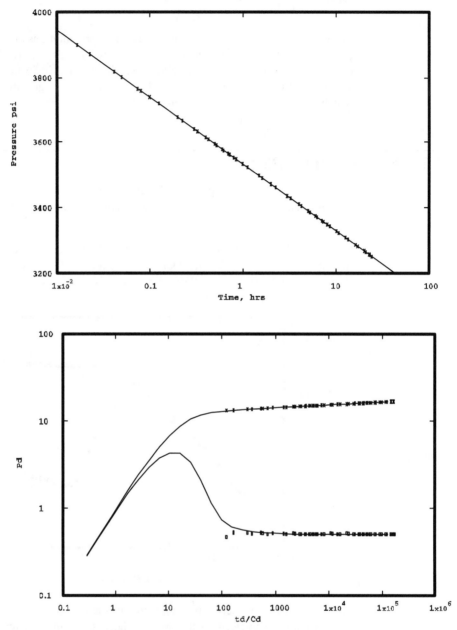

Figure 3.2: Infinite acting radial flow shows as semilog straight line on a semilog plot, as a flat region on a derivative plot

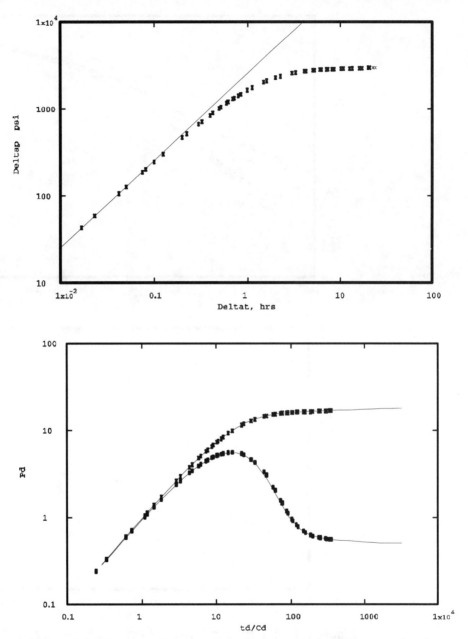

Figure 3.3: Storage shows as a unit slope straight line on a log-log plot, as a unit slope line plus a hump on a derivative plot. The "hump" on the derivative plot is characteristic of a damaged well (positive skin).

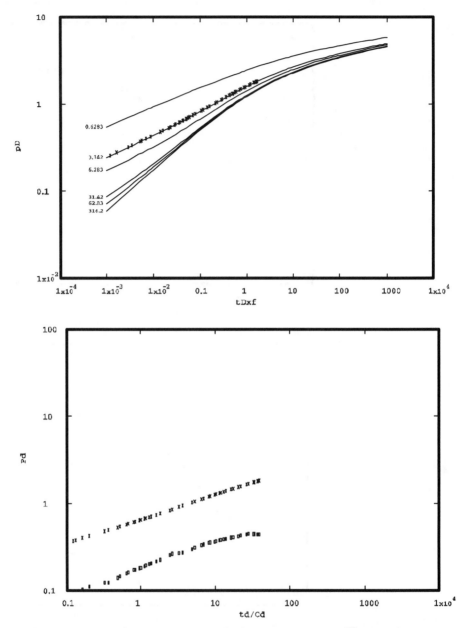

Figure 3.4: A finite conductivity fracture shows as a $^1/_4$ slope line on a log-log plot, same on a derivative plot. Separation between pressure and derivative is a factor of 4.

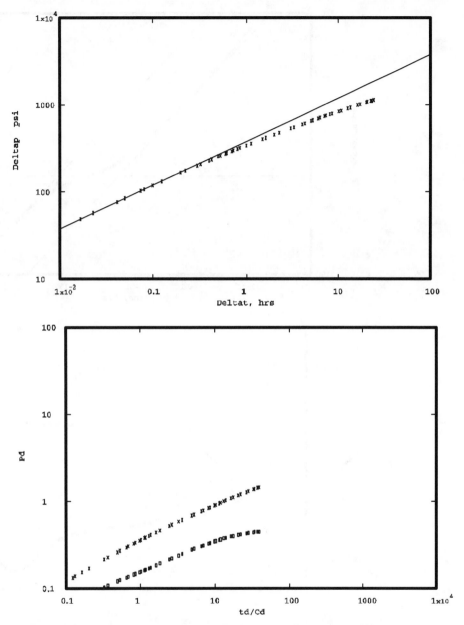

Figure 3.5: An infinite conductivity fracture shows as a ½ slope line on a log-log plot, same on a derivative plot. Separation between pressure and derivative is a factor of 2.

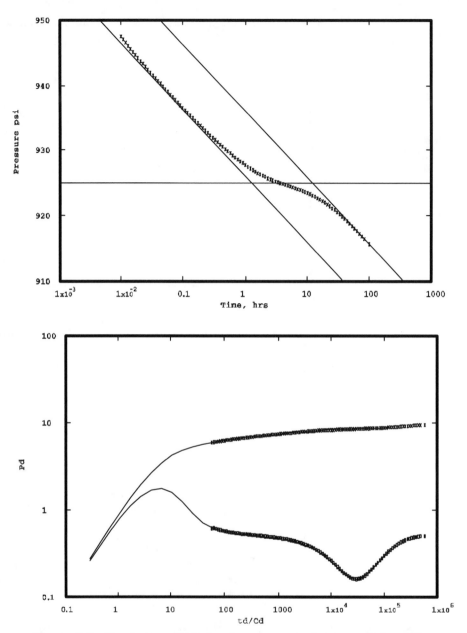

Figure 3.6: Dual porosity behavior shows as two parallel semilog straight line on a semilog plot, as a minimum on a derivative plot.

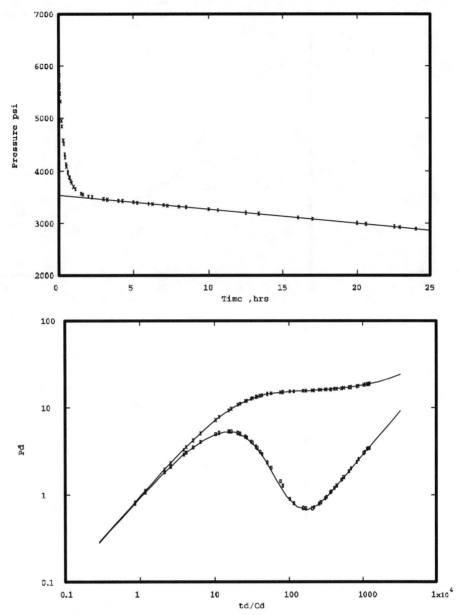

Figure 3.7: A closed outer boundary (pseudosteady state) shows as a straight line on a Cartesian plot, as a steep rising straight line of unit slope on a derivative plot.

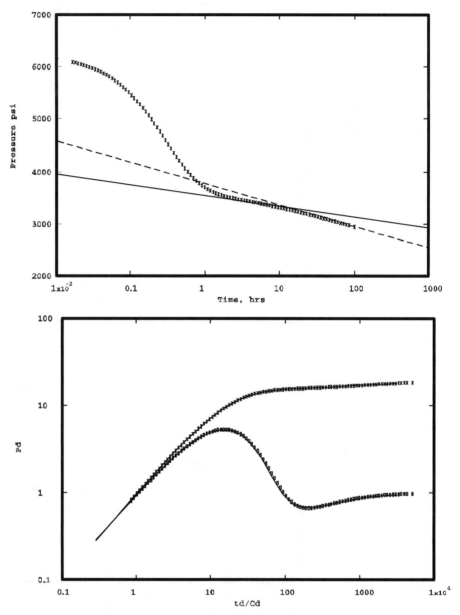

Figure 3.8: A linear impermeable boundary shows as semilog straight line with a doubling of slope on a semilog plot, as a second flat region on a derivative plot.

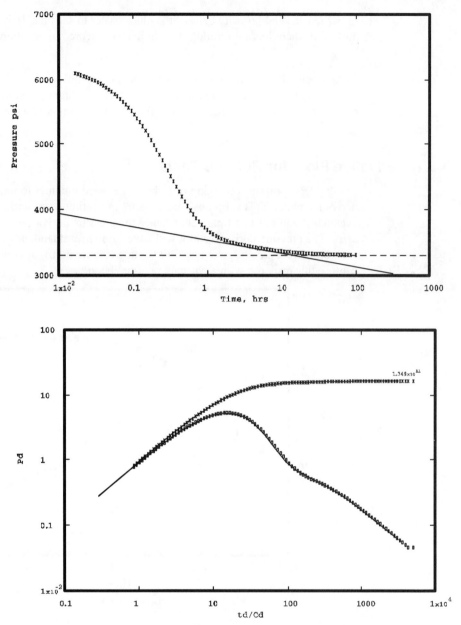

Figure 3.9: A constant pressure boundary shows as flat region on most p vs. t plots, as a continuously decreasing line on a derivative plot.

In particular, dual porosity behavior is much easier to see on a derivative plot (Fig. 3.6) even when the first semilog straight line is obscured by wellbore storage.

Even though the derivative plot is by far the most useful for diagnosis, it is not necessarily the most accurate for calculation when it comes to estimating parameters. Hence the other plots (particularly the semilog plots) are still required.

3.3.1 Derivative Plots for Buildup Tests

Buildup tests present something of a challenge when it comes to interpreting derivative plots. This is because the pressure in a buildup test will always eventually stabilize to a final value, hence the pressure derivative trends towards zero. The pressure derivative plot will have a downward tendency at later time, even though the flow regime is infinite-acting (Figure 3.10). It is important not to confuse this response with a constant pressure boundary.

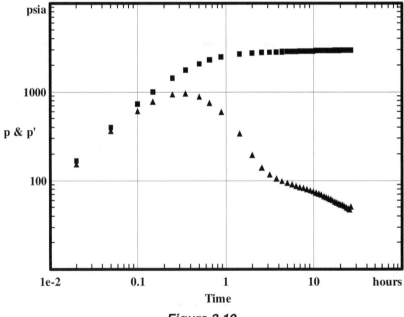

Figure 3.10

As described in "2.15.2 Treating Buildups as Drawdowns" on page 60, buildup tests can often be treated in the same manner as drawdown tests, provided that the pressure transient is in infinite-acting behavior and provided that a suitable time variable is used (for example, the Horner time or Agarwal effective time). Since the derivative plot uses a log-log scale, the effective time can be used to "straighten" the flat region representing infinite-acting radial flow, as shown in Fig. 3.11 which can be compared to Fig. 3.10.

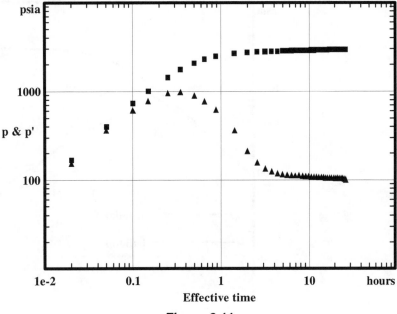

Figure 3.11

The effective time can only straighten the derivative plot when both the buildup period and the drawdown period that underlies it are both exhibiting semilog behavior (infinite-acting radial flow). If the reservoir is responding to some kind of boundary, then the derivative plot is likely to have a downward trend even for pseudosteady state behavior. Fig. 3.12 compares the buildup and drawdown responses for the same closed reservoir (pseudosteady state behavior) -- the drawdown behavior has an upward trending derivative, whereas the buildup behavior has a downward trending derivative even when the effective time is used. Fig. 3.13 makes a similar comparison for drawdown and buildup responses for a reservoir with a constant pressure boundary -- in this case both responses have the downward trending derivative.

In summary, the diagnosis of the derivative plot for a buildup test requires some care. If the producing time has been relatively short, the derivative can be expected to have a downward trend at late time. If the reservoir response is infinite acting, the downward trend can be removed by using effective time (although it should be noted that this will not work properly unless the correct producing time is used). If the downward trend is not removed by use of effective time, then there is probably some kind of boundary effect -- exactly which kind of boundary effect is not always clear, since all types of boundary affect the derivative in roughly the same manner. Although desuperposition can sometimes be helpful, the best solution to this difficulty probably lies in the geological information about the reservoir -- in other words, the interpreter needs to have a pretty good idea what kind of boundary to be looking for in advance.

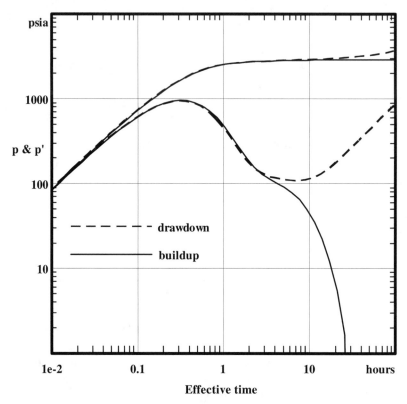

psia

p & p'

1000

100

10

- - - - - drawdown

——— buildup

1e-2   0.1   1   10   hours

Effective time

Figure 3.12

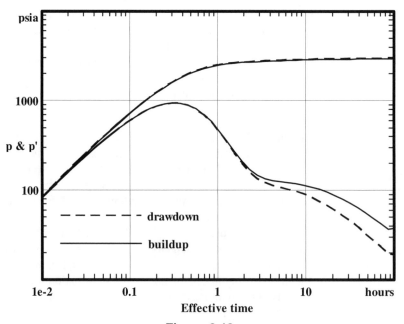

psia

p & p'

1000

100

- - - - - drawdown

——— buildup

1e-2   0.1   1   10   hours

Effective time

Figure 3.13

3.3.2 Calculating Derivatives

Calculating the pressure derivative requires some care, since the process of differentiating the data amplifies any noise that may be present. A straightforward numerical differentiation using adjacent points (Eq. 3.1) will produce a very noisy derivative (Fig. 3.14).

$$t\left(\frac{\partial p}{\partial t}\right)_i = t_i \left[\frac{(t_i - t_{i-1})\Delta p_{i+1}}{(t_{i+1} - t_i)(t_{i+1} - t_{i-1})} \right.$$

$$\left. + \frac{(t_{i+1} + t_{i-1} - 2t_i)\Delta p_i}{(t_{i+1} - t_i)(t_i - t_{i-1})} - \frac{(t_{i+1} - t_i)\Delta p_{i-1}}{(t_i - t_{i-1})(t_{i+1} - t_{i-1})} \right] \qquad (3.1)$$

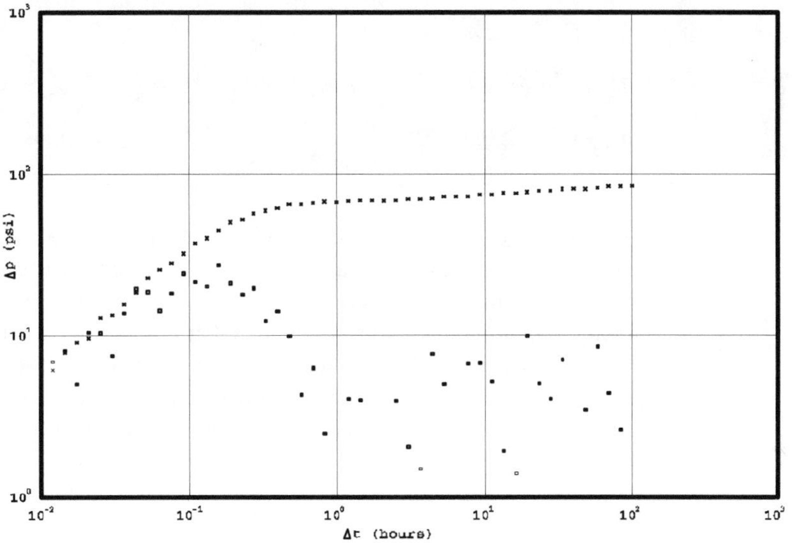

Figure 3.14

If the data are distributed in geometric progression (with the time difference from one point to the next becoming greater as the test proceeds), then the noise in the derivative can be reduced somewhat by using a numerical differentiation with respect to the logarithm of time:

$$t\left(\frac{\partial p}{\partial t}\right)_i = \left(\frac{\partial p}{\partial \ln t}\right)_i$$

$$= \left[\frac{\ln(t_i / t_{i-1})\Delta p_{i+1}}{\ln(t_{i+1} / t_i)\ln(t_{i+1} / t_{i-1})} + \frac{\ln(t_{i+1} t_{i-1} / t_i^2)\Delta p_i}{\ln(t_{i+1} / t_i)\ln(t_i / t_{i-1})} \right.$$

$$\left. - \frac{\ln(t_{i+1} / t_i)\Delta p_{i-1}}{\ln(t_i / t_{i-1})\ln(t_{i+1} / t_{i-1})} \right] \qquad (3.2)$$

However, even this approach leads to a noisy derivative. The best method to reduce the noise is to use data points that are separated by at least 0.2 of a log cycle, rather than points that are immediately adjacent. Hence:

$$t\left(\frac{\partial p}{\partial t}\right)_i = \left(\frac{\partial p}{\partial \ln t}\right)_i$$

$$= \left[\frac{\ln(t_i / t_{i-k})\Delta p_{i+j}}{\ln(t_{i+j} / t_i)\ln(t_{i+j} / t_{i-k})} + \frac{\ln(t_{i+j}t_{i-k} / t_i^2)\Delta p_i}{\ln(t_{i+j} / t_i)\ln(t_i / t_{i-k})} \right.$$

$$\left. - \frac{\ln(t_{i+j} / t_i)\Delta p_{i-k}}{\ln(t_i / t_{i-k})\ln(t_{i+j} / t_{i-k})} \right]$$

(3.3)

$$\ln t_{i+j} - \ln t_i \geq 0.2$$

(3.4)

$$\ln t_i - \ln t_{i-k} \geq 0.2$$

(3.5)

The value of 0.2 (known as the **differentiation interval**) could be replaced by smaller or larger values (usually between 0.1 and 0.5), with consequent differences in the smoothing of the noise. Figs. 3.15(a) through (c) compare the different amount of smoothing achieved. Notice that if a very wide interval is used (0.5 in Fig. 3.15c), then the shape of the calculated derivative curve (represented by the points in Figs. 3.15a-c) may be distorted. In Fig. 3.15(c), the points to the right of the storage "hump" are displaced to the right, by comparison with Figs. 3.15(a) and (b).

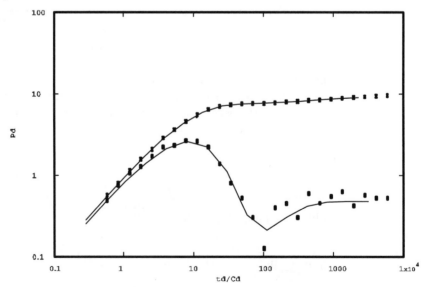

Figure 3.15(a): Differentiation interval 0.1

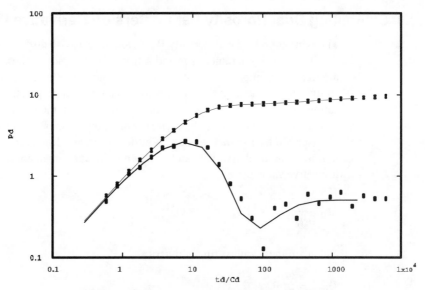

Figure 3.15(b): Differentiation interval 0.2

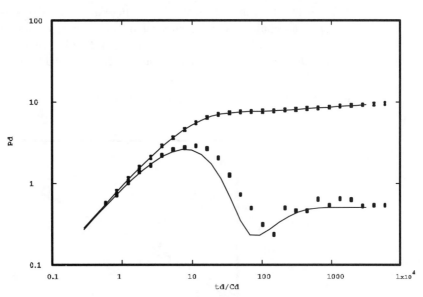

Figure 3.15(c): Differentiation interval 0.5

It should be noted that the use of the differentiation interval may cause problems determining the derivative in the last part of the derivative curve, since the data "runs out" within the last differentiation interval. Some noise is therefore to be expected at the end of the data. Also, the differentiation interval approach may overly flatten the early time derivative, and since this part of the data is not prone to noise anyway, it is often better to use the arithmetic differentiation for the early points (Eq. 3.2).

3.3.3 Estimating Dual Porosity Parameters on Derivative Plots

The section "2.11 Dual Porosity Behavior" on page 41 discussed the estimation of the dual porosity parameters ω and λ from semilog plots. These semilog methods are frequently impractical because wellbore storage effects hide the earlier semilog straight line. The derivative plot provides a much more practical method for estimating ω and λ. The position of the minimum in the derivative (the "dip" that characterizes dual porosity behavior) completely defines the values of both ω and λ as described by Bourdet, Ayoub, Whittle, Pirard and Kniazeff (1983). The minimum in the derivative can be shown to lie at a dimensionless pressure derivative value of:

$$\left[t_D \frac{\partial p_D}{\partial t_D} \right]_{min} = \frac{1}{2}\left[1 + \omega^{\frac{1}{1-\omega}} - \omega^{\frac{\omega}{1-\omega}} \right]$$

(3.6)

and at a dimensionless time value of:

$$t_{Dmin} = \frac{\omega}{\lambda}\ln\frac{1}{\omega}$$

(3.7)

The value of ω can be evaluated by solving Eq. 3.6 iteratively using a Newton-Raphson technique, after which λ can be found directly from Eq. 3.7. The iterative determination of ω can be made much more robust by providing a good first estimate of its value. Such a first estimate can be found from:

$$\log\omega \approx \left(0.01765 + \log\left[t_D \frac{\partial p_D}{\partial t_D} \right]_{min} \right) / 0.94903$$

(3.8)

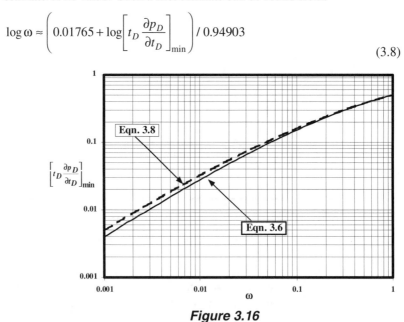

Figure 3.16

Figure 3.16 shows a comparison between Eqs. 3.6 and 3.8, and can be used directly to estimate ω if required.

As an example of this procedure, Fig. 3.17 shows the derivative plot from a well test in a dual porosity reservoir. The location of the minimum is at a pressure value of 1.02 psi and a time value of 7 hrs. Using known values of q, B, μ and h, together with a previously estimated value of k, the dimensionless pressure derivative at the minimum can be calculated, after which the first estimate of ω is found from Eq. 3.8 and a final estimate of ω is found by Newton-Raphson solution of Eq. 3.6. After estimating ω, the dimensionless time value at the minimum can be calculated using known values of ϕ, μ, c_t and r_w together with the estimated value of k, after which λ can be estimated using Eq. 3.7 (which also requires the inclusion of the recent estimate of ω).

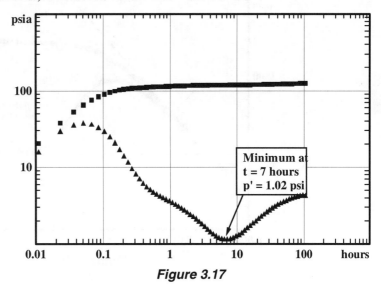

Figure 3.17

3.4 Diagnostic Plot Evaluation

As has been described in the previous sections, different parts of the reservoir response are recognizable by their characteristics or particular graphical presentations. This enables the engineer to separate one part of the response from another. It is absolutely critical to the final interpretation that this distinction be made. Why is it so important? Consider, for example, the estimation of reservoir permeability from the slope of the semilog straight line characteristic of infinite acting radial flow -- there may be other parts of the response which may at first appear to show a semilog straight line, but which would give totally erroneous estimates of the permeability. Such false straight lines could be due to boundary effects, or may be due to external effects completely unrelated to the reservoir response (for example temperature response of the instrument, removal of drilling fluids from the invaded zone, etc.). Since certain specific portions of the response are used to estimate specific reservoir parameters, it is clearly necessary to identify each portion precisely.

Often a good indication of a particular reservoir response can be obtained by considering the responses preceding and following it, since the different responses do come in a certain chronological order (this was summarized in Table 2.1). Thus we would not look for infinite acting radial flow before wellbore storage, nor would we look for it after a pseudosteady state response. It is often useful to "tag" particular response regions (e.g., storage, semilog straight line, boundary effect) to confirm that the identified responses appear in the correct order, and do not overlap each other (Fig. 3.18).

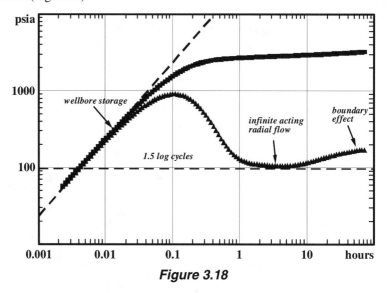

Figure 3.18

There are also useful indicators in some of the transitions between flow regimes, for example, the 1½ log cycles between storage and infinite acting radial flow, which gives rise to the **1½ log cycle rule**. The specific response characteristics have already been illustrated in Figs. 3.2 - 3.9, and will be discussed in more detail in specific examples later.

3.5 Data Preparation

If we are to be able to correctly identify regions of the data, it is clearly important that the data itself be correct. Since real data collection is sometimes subject to both physical and human error, it is worthwhile to consider the preparation of data prior to beginning the analysis. Some preparation may be necessary to extract the pressure data in the correct format, units and datum from service company output files, however this is relatively mundane. Two specific items of data preparation that require special care are discussed in detail here, namely (a) number and frequency of data, and (b) datum shifting.

3.5.1 Number and Frequency of Data

Modern electronic pressure gauges are capable of high precision measurements at high frequency, making digital pressure measurements faster than one per second if required. These measurements can be recorded at surface (in surface recording tools) or may be stored electronically within the downhole tool itself (in memory tools). In either case, the total number of data points can be many thousands during the time of a test. In permanently installed gauges, data can accumulate to millions of points. Plotting and manipulating so many data can be a monstrous task, even for a computer, and it is usually desirable to reduce the total data set to a more manageable size by sampling a representative subset of at most a few hundred data.

The main objectives in sampling data are: (a) to capture essential features of the response; (b) to apply roughly equal weight to each of the separate response regions; (c) to reduce the total number of data to speed up plotting and calculation.

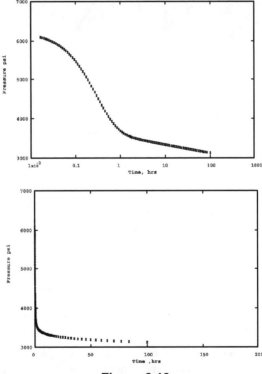

Figure 3.19

If our objective were only to reduce the number of data, it would be legitimate to sample data at fixed intervals (for example sampling every tenth point in 2000 data, to end up with 200 data). However, if the data were measured at regular intervals, this would tend to provide an excessive number of late time points and too few early time points, since we know that the most rapid changes take place at early time.

This kind of sampling can be useful for preliminary reduction of very large data sets, or when the data were actually recorded with reducing frequency at later time.

If data were recorded at a uniform spacing in time, a more suitable sampling would be *logarithmic*. This approach includes more data at early time and fewer at late time (Fig. 3.19). This achieves the correct weighting for infinite acting responses, which we have already observed are logarithmically dependent on time. However this would not be a good method to use when the data includes several different flow periods, since the logarithmic cycle needs to be restarted at each major flow change -- Fig. 3.20 shows an example of a logarithmic sampling in a multirate test, demonstrating that insufficient data points are included after the rate change.

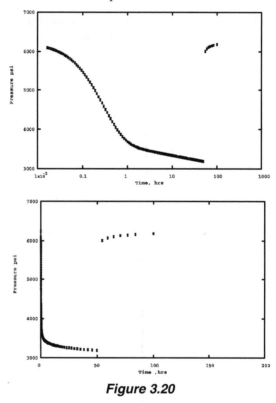

Figure 3.20

A third sampling method that is more generally applicable is one which samples uniformly in *pressure* rather than in time. This ensures that more data are included at times where the pressure changes most, which will usually result in more or less equivalent weighting of all flow regimes. Care must be taken in specific cases. Firstly, if the data contains flat regions (regions of constant pressure) then it is necessary to ensure that the region is not skipped altogether. Secondly, if the data is noisy in some places relative to others, then this sampling method will tend to emphasize the noisier data.

3.5.2 Datum Shifting

An instrument may measure pressures at one second intervals, and may record during a well test for several days. Hence data may be available from 1 second to 10^5 or 10^6 seconds, and may span six log cycles in a graphical presentation. However it must be remembered that a logarithmic presentation tends to hide errors at late time, and emphasize them at early time. It must also be remembered that the start of the test does not occur instantaneously, nor does the pressure gauge react immediately. Opening or closing a wellhead valve may take some minutes, and the pressure transducer may take up to a minute to stabilize (this is particularly true of capillary tube tools). For these various reasons, the actual time of the early data may be a minute or more different from the recorded time. Another way of looking at the situation is that at the indicated time, the pressure was not really that recorded at that same actual time. Errors of a few minutes in time, or a few psi in pressure drop may not seem significant, but because of the common use of logarithmic plots such differences can substantially alter the appearance of the data and can therefore impair the interpretation. Consider the effects of one minute changes in the specification of time zero in Figure 3.21.

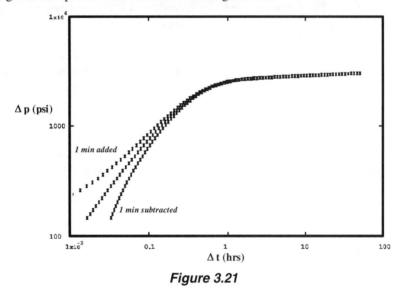

Figure 3.21

In this example it would be easy to mistake the wellbore storage region for some kind of fracture effect (in the case of the upper curve).

Thus it is important to be cautious with early time data, especially from modern electronic gauges. It may often be necessary to shift the time datum forward or backward to correct the appearance of the early time data. Another way of correcting the situation is to treat the initial pressure measured at the start of the test as a variable (this would be p_i for a drawdown test, or p_{wf} for a buildup test). This will be discussed in more detail in Section 3.6.2.

3.6 Nonlinear Regression

One of the most powerful analytic tools made possible by computer-aided interpretation is **nonlinear regression** (sometimes known as **automated type curve matching**).

This method is entirely different from graphical techniques in that it uses a mathematical algorithm to match the observed data to a chosen reservoir model. The matching is achieved by changing the values of the unknown reservoir parameters (such as permeability, skin, ω, λ, distance to boundary, etc.) until the model and the data fit as closely as possible (in a least squares sense) by minimizing the sum of squares of the differences between measured pressure and model pressure:

$$\sum_{i=1}^{n}\left[p_{measured}(t_i) - p_{model}(t_i, k, s, C, ...)\right]^2$$

The nonlinear regression procedure is freed from the restrictions associated with graphical techniques (constant rate production and instantaneous shut-in) and can therefore be used to interpret more modern tests in which downhole rates were recorded or in which several different rates were used. The method fits all the data simultaneously, and therefore avoids the problem of inconsistently interpreting separate portions of the data, as can happen with graphical analysis. Also, the mathematical fitting process allows for the statistical determination of goodness of fit, thus providing not only a numerical answer but also a quantitative evaluation of how good the answer is. Finally, nonlinear regression is capable of estimating reservoir parameters from pressure responses that are in the transition regions which cannot be interpreted directly by graphical methods. Thus there are significant cases of tests which are interpretable by nonlinear regression but not by graphical techniques, such as those that terminated prior to reaching the semilog straight line.

To summarize the advantages of using nonlinear regression, they are:

 (a) Analyzes multirate or variable rate tests,

 (b) Avoids inconsistent interpretations,

 (c) Provides confidence estimate on answer,

 (d) Can interpret "uninterpretable" tests.

At the same time, nonlinear regression requires the specification of the reservoir model to be matched; the algorithm itself does not select which model is appropriate. Thus a nonlinear regression analysis must be complemented by a visual diagnosis of the data so that the engineer can select the correct reservoir model. At the same time as selecting the reservoir model the engineer can also speed up the nonlinear regression by making first estimates of the reservoir parameters. Most computer-aided well test software will do this automatically.

3.6.1 Confidence Intervals

To understand the evaluation of statistical goodness of fit, it is useful to discuss the concept of confidence intervals. Suppose that we estimate the reservoir permeability by fitting a semilog straight line through data from a pressure transient test, as in Fig. 3.22.

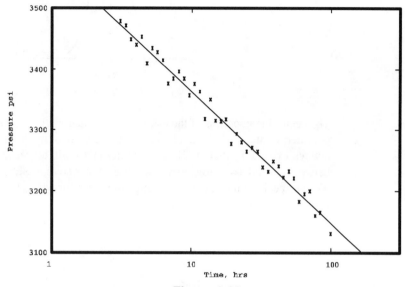

Figure 3.22

Since the data do not actually lie exactly in any straight line, there is some ***uncertainty*** as to whether the straight line is the correct one or not. If several people perform the fit graphically, then they are likely to obtain slightly different answers. Using regression, the answer will always be the same, but since the data are noisy perhaps under a different realization of the noise the answer would have been different. Hence we can say that although our answer is the most likely, due to the uncertainty in the data there is a possibility that the true answer is different. There is high probability that the true answer is close to our best estimate, and low probability that it is far away. If we assume that the probability of a given answer is normally distributed about our best estimate, then we can recognize a "good" answer from a "bad" answer by the shape of the distribution (Fig. 3.23).

In Fig. 3.23(a), the probability that the true answer is close to x^* is high, and the probability that the answer is far from x^* is very small. Hence x^* is a "good" estimate. On the other hand, in Fig. 3.23(b), the true answer is almost as likely to have a value quite far from x^* as it is to have the value x^* itself. Hence x^* is a "bad" estimate in this case.

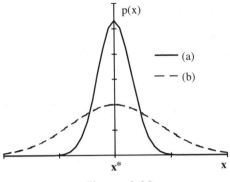

Figure 3.23

To specify the goodness of the estimate more quantitatively, we can define the confidence interval. The total area under the probability distribution curve is always 1.0, that is, there is 100% certainty that x will take a value somewhere between $-\infty$ and $+\infty$. However, if we restrict ourselves to **95% confidence**, then we can specify a much narrower range of values for which the included area under the curve is 0.95, as in Fig. 3.24.

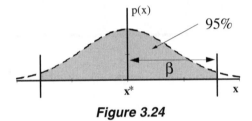

Figure 3.24

In this case we can say with 95% confidence of being right that the true value of x lies within the range:

$$x^{*}-\beta < x < x^{*}+\beta \tag{3.9}$$

This range of values is known as the **95% confidence interval** on the estimate value of x. There are various ways of specifying an answer together with its confidence interval, one would be to specify in terms of the absolute value of the range, for example $x^{*}\pm\beta$, while another would be to specify the range in terms of a percentage of the value of the parameter, for example $x^{*}\pm(100\,\beta/x^{*})$ percent.

When using 95% confidence intervals to evaluate the goodness of fit of a nonlinear regression analysis of well tests, it is useful to define an **acceptable estimate** as one with a confidence interval that is 10% of the value itself. This definition is not useful for skin factor (since it can have value zero) or for initial reservoir pressure (since it can have large values). Table 3.2 summarizes the suggested limits on acceptable confidence ranges. The values in this table are based on a match of *pressure* -- when matching pressure *derivative* the acceptable confidence intervals can be roughly twice as wide.

Parameter	Acceptable % interval	Acceptable absolute interval
k	10%	
C	10%	
ω	20%	
λ	20%	
r_e	10%	
x_f	10%	
s		1.0
p_i		1.0 (psi)

Table 3.2

The confidence interval is a function of the noise in the data, the number of data points, and the degree of correlation between the unknowns. As such, it is a useful indicator of whether the data is too noisy to confidently support the estimate obtained, and is also a good indicator of ambiguity (for example if the same fit could be obtained by changing the permeability estimate and making a compensating change in the skin estimate). No such indicators are possible with traditional graphical analysis, and in the past it was not possible to be assured as to how valid a given estimate was.

In simple terms, there is reason to question the validity of a reservoir model if *any* of its parameters have confidence intervals that are wider than the acceptable range, however in practice, a less stringent condition may be used. If data are missing over a given flow regime (e.g. wellbore storage), then the confidence interval for this region will be wide. However, the answers that originate from the other flow regions (for which there are data) may have confidence intervals within the allowable range, and their estimated values are still acceptable. Figs. 3.25(a) and (b) illustrate this point -- removing the wellbore storage dominated portion of the data results in a substantial widening of the confidence interval on the estimate of the wellbore storage parameter (Fig. 3.25b). The permeability and skin estimates are unchanged and their confidence intervals widen only somewhat since there are fewer data points in Fig. 3.25(b).

On the other hand, matching a reservoir model that is inappropriate for the data should give confidence intervals that are unacceptable for most (or all) of the estimates, such as in Fig. 3.26.

Confidence intervals are also useful indicators that an apparently good match may in fact have insufficient data to be of significance (Fig. 3.27).

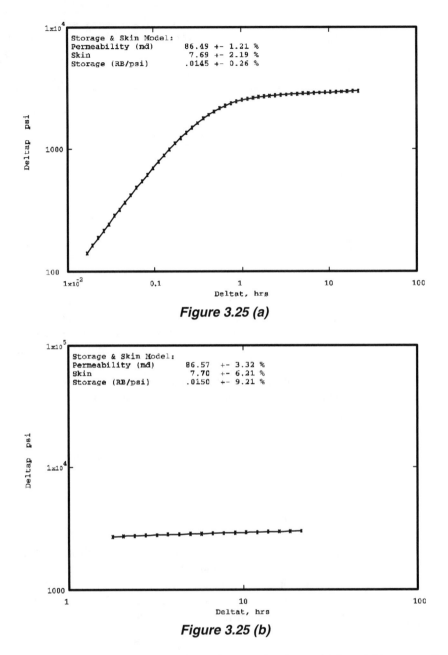

Figure 3.25 (a)

Figure 3.25 (b)

Examination of confidence intervals teaches us a lot about which variables are well determined in a well test, and which are not. For example, permeability can be found with a very high degree of confidence from a test that contains a long period of infinite acting radial flow (semilog straight line), but the dual porosity parameters ω and λ are always much more indefinite, even when matching using pressure derivative (which is more sensitive to ω and λ than pressure itself).

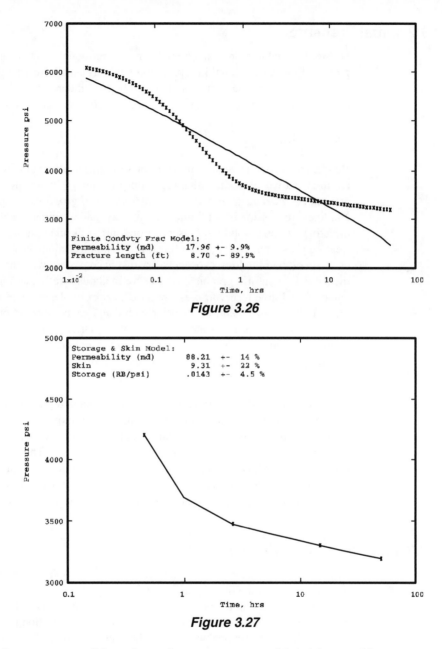

Figure 3.26

Figure 3.27

In summary, confidence intervals represent a powerful tool that provides interpretation information that is not available in traditional graphical well test analysis. To be able to discover the degree of significance of the answer is one of the primary advantages of computer-aided well test interpretation.

3.6.2 Initial Pressure

Due to the need to represent a complex reservoir model in a two-dimensional graphical form, traditional interpretation methods usually restrict the number of parameters that can be estimated, and fix the remainder. In particular, most type curves are based on representations of pressure **drop** rather than pressure itself:

$$\Delta p = p_i - p = \frac{141.2 q B \mu}{kh} p_D(t_D)$$

(3.10)

Hence, the need to present a single graph of $p_D(t_D)$ to allow graphical analysis requires that we know the initial reservoir pressure, p_i, in advance. *** This is an arbitrary requirement forced by the restrictions of graphical interpretation.*** Nonlinear regression is not limited in this way, and is able to match pressure directly by treating the initial reservoir pressure p_i as an unknown variable. By treating p_i as an unknown, a nonlinear regression technique is able to make an estimate of the initial reservoir pressure that is consistent with the entire range of the data. By contrast, graphical type curve methods force an assumption of the value of the initial reservoir pressure based, for example, on the early time measurements -- as mentioned earlier such data may be in error, or may not represent an initial stable state in the reservoir. These assumptions in the graphical approach reduce the precision of estimates of initial pressure, and may also adversely affect the other reservoir parameter estimates.

Since computer-aided interpretation often uses standard graphical presentations for preliminary diagnosis and model selection, it is common to include standard approximations for the initial pressure (such as pressure just prior to opening for a drawdown test, or just prior to shut-in for a buildup test). These should normally be used for the graphical presentations only, and except in rare instances, it is almost always desirable to **release** the assumption of initial pressure during nonlinear regression. Initial pressure is often a parameter that is usefully learned in a well test.

3.6.3 Multirate and Variable Rate Tests

In the same way that graphical analysis techniques require prior knowledge or assumption about the initial pressure value, they are similarly restricted to **constant rate** tests. Other than the practical necessity of maintaining a flow rate that can be registered, this limitation to constant rate is another restriction imposed by the necessities of the graphical techniques themselves. The entire concept of wellbore storage is required only to accommodate the fact that the sandface flow rate is not constant during the early part of the test. Estimating the wellbore storage coefficient has no inherent value of its own, and if the actual sandface flow rate could be measured and used in the analysis, then there would be no need to introduce wellbore storage into the problem at all. Modern downhole instruments have made reliable flow rate measurement a reality, and nonlinear regression provides a method of interpreting such information without imposing the constant rate restriction required for most graphical techniques. The same approach can also be used to include the effects of a series of different (constant) flow rates prior to or during the well test (Fig. 3.28). Such a "multirate" test would not require the use of a downhole flow meter, but the wellbore storage effect would always need to be included if such downhole data were not available. The pressure change during such a multirate test with nq different rates can be represented by superposition in time:

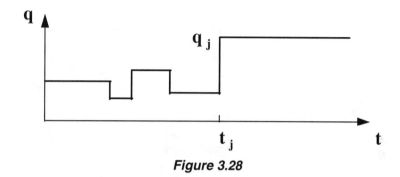

Figure 3.28

$$p(t) = p_i - \frac{141.2qB\mu}{kh} \sum_{j=1}^{nq}\left(q_j - q_{j-1}\right)p_D(t_D), \quad t > t_j$$

(3.11)

where t_j is the time at which flow step q_j started. Continuously varying flow rate data can be used in exactly the same manner, treating each measured data point as a small "stairstep" of constant rate.

Nonlinear regression can accommodate such varying rate data just as easily as a constant rate test (although calculating the summation may take longer). In fact, incorporation of variable rate includes more information about the reservoir. Why is this so? Every measured pressure point contains ***early time*** reservoir response from the most recent flow rate change and ***later time*** response from earlier flow rate changes. Thus the reservoir pressure signal contains more indicators of early time behavior and late time behavior than a test conducted at constant rate (which has only one early time response and one late time response). Consider the example of a buildup test shown in Fig. 3.29 in which the well was shut in briefly to position the pressure bomb prior to the final closure.

Traditional analysis would probably assume the pressure to be stabilized prior to the final shut-in, and would ignore the short closure. Performing traditional rate normalization would be difficult due to the period of zero flow rate. However, inclusion of the actual flow rate history is not only more accurate, it includes acknowledgment of the pressure response due to the earlier closure. Thus instead of the effective time of the test being Δt, it can in fact be considered to be $\Delta t + (t_p - t_1)$, even though pressure measurements did not start until time t_p. The effective length of the test is therefore longer than the actual measurement time that would have been analyzed (imprecisely) by traditional graphical techniques.

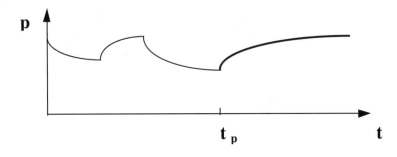

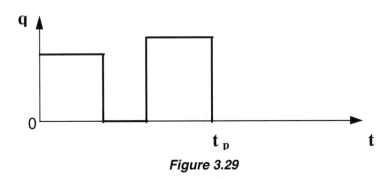

Figure 3.29

Consider a second example in which the flow rate was measured downhole in a gas well during shut-in (Fig. 3.30). Fig. 3.30 shows the real data that was presented by Guillot and Horne (1986). Due to wellbore storage effects, the sandface flow rate did not reach zero until about 20 minutes after the wellhead valve closed. At this time (20 min.) the reservoir was flowing at such a low rate that the flow rate tool (spinner) stopped rotating and indicated zero flow. Using nonlinear regression, together with the downhole flow rate data, it was possible to estimate the reservoir properties from only this 20 minutes of data, even though wellbore storage dominated this entire period (and later).

This is a particularly significant true life example in that it demonstrates how the use of downhole flow rate measurements and nonlinear regression made it possible to analyze 20 minutes of data in a test that would need to run for 8 hours to be interpretable by traditional techniques. Furthermore, using the same technique in a gas well test, Horne and Kuchuk (1988) were able to estimate rate-dependent skin effect and absolute open flow potential (*AOFP*) from a buildup test, even though traditional methodology requires a complex and time consuming isochronal test to determine these parameters. It is possible to estimate the rate-dependent skin effect since the reservoir is actually flowing into the wellbore throughout the afterflow (storage dominated) period. Having estimated the rate-dependent skin, it is straightforward to draw a deliverability diagram using the measured rate data and to extrapolate to determine the *AOFP*.

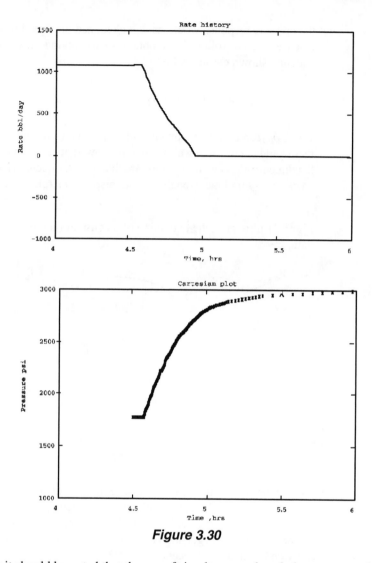

Figure 3.30

Finally it should be noted that the use of simultaneous downhole pressure and flowrate measurements is unaffected by any complex wellbore phenomena such as changing wellbore storage or phase segregation ("humping effects"). As flow rate tools become more highly developed for use in multiphase environments, these techniques should become widely used.

3.7 Desuperposition

Despite the advantages of variable rate analysis illustrated in the previous section, an obvious difficulty is that it is no longer possible to perform diagnosis by examining standard plots, since the familiar characteristics may not appear. For the purposes of diagnosis, a process of **desuperposition** is used to calculate the way

---

the pressure response would have looked, had the flow rate been constant. The essence of deconvolution is to obtain p_{0D} by solving the convolution integral equation shown earlier as Eq. 2.58:

$$p_{wD} = \int_0^{t_D} q'_D(\tau)\left[p_D(t_D - \tau) + s\right]d\tau \qquad (3.12)$$

where p_{0D} represents the dimensionless reservoir pressure response (without skin) that would occur if the well were to be flowed at constant rate, and p_{wD} represents the dimensionless wellbore pressure that actually occurred due to the flow rate q_D, where q_D is nondimensionalized with respect to a reference flow rate $q_D=q/q_{ref}$.

Fig. 3.31 illustrates the extraction of a constant rate response from variable rate data.

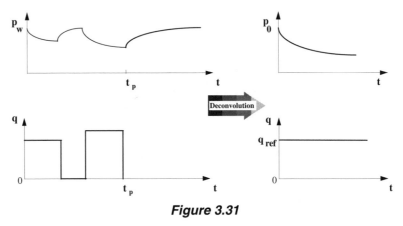

Figure 3.31

Techniques of desuperposition require mathematical techniques that will not be described in detail here, but include constrained optimization as in Kuchuk and Ayestaran (1985) and Laplace transform techniques as in Roumboutsos and Stewart (1988). Mendes, Tygel and Correa (1989), Kuchuk (1990) and Raghavan (1993) provide summaries of recent approaches.

Despite the attention placed on the subject, rigorous and robust deconvolution procedures have proven to be elusive due to numerical instabilities inherent in the inversion of the convolution integral. One traditional deconvolution method that has recently gained renewed favor is the **rate-normalization** procedure described originally for buildup tests by Gladfelter, Tracy and Wilsey (1955) and by Winestock and Colpitts (1965) for drawdown tests.

The procedure normally known as rate-normalization consists of a simple approximation of the convolution integral (Eq. 3.12):

$$p_{wD}(t_D) \approx q_D(t_D)p_{0D}(t_D) \qquad (3.13)$$

Using this simple approximation, it is straightforward to find the constant rate response by rearranging the equation:

$$p_{0D}(t_D) \approx p_{wD}(t_D) / q_D(t_D)$$

(3.14)

Hence the basis of rate-convolution is to plot $[p_i - p_{wf}(t)]/q(t)$ vs. t in the case of a drawdown test or $[p_{ws}(\Delta t) - p_{wf}(t_p)]/[q(t_p) - q(t_p + \Delta t)]$ vs. Δt in the case of a buildup test. Obviously, there is an error inherent in making this simplification of the convolution integral, and the magnitude and form of the error can be determined as shown in Sections 9.3 and 9.4 of Raghavan (1993). Nonetheless, from a practical standpoint the rate-normalization procedure is sufficiently accurate provided the slope of the rate-time plot on log-log coordinates is less than 0.1. Rate-normalization is also very attractive from the point of view of ease of computation. It should always be remembered that the most important use of deconvolution is to aid in the identification of the reservoir model, rather than for estimation of reservoir parameters (which can be done without approximation using nonlinear regression on the original [not deconvolved] data, once the reservoir model has been identified).

Given its primary role as an aid to model recognition, one useful new method is the **Laplace pressure** approach described by Bourgeois and Horne (1993). Laplace transform deconvolution techniques such as those described by Roumboutsos and Stewart (1988) and Mendes, Tygel and Correa (1989) make use of the powerful property that the Laplace transform of the convolution integral is a simple product. Hence, in Laplace space:

$$\overline{p}_{wD}(z) = \overline{q}_D(z)\overline{p}_{0D}(z)$$

(3.15)

where the overbar represents transformation with respect to t_D, and z is the Laplace variable. Unlike the rate-normalization simplification in Eq. 3.13, there is no approximation in Eq. 3.15. Hence the constant rate pressure can be found exactly in Laplace space as:

$$\overline{p}_{0D}(z) = \overline{p}_{wD}(z) / \overline{q}_D(z)$$

(3.16)

The basis of many Laplace based deconvolution approaches is the inversion of the transform of $\overline{p}_{0D}$, however the numerical instabilities inherent in this process often cause difficulty. Bourgeois and Horne (1993) suggested plotting variables in Laplace space, using the **Laplace pressure** $z\overline{p}(z)$ as the pressure variable and $1/z$ as the time variable. Such a plot, and its corresponding derivative, is very similar in nature to a conventional plot of pressure vs. time and is therefore very useful for model recognition.

As an example of the use of Laplace pressure and deconvolution, consider the variable rate drawdown test shown in Fig. 3.32. These test data were shown in Kuchuk (1990). Due to an unexpectedly large pressure drop, the well was in risk of passing below the bubble-point pressure and was therefore throttled back. The resulting pressure response showed an increasing pressure and a consequently strange looking derivative plot (Fig. 3.33). The deconvolved Laplace pressure plot in Fig. 3.34 shows a much more conventional looking response, after which nonlinear regression can be performed using the appropriate model (the nonlinear regression match to the original data is shown in Fig. 3.33).

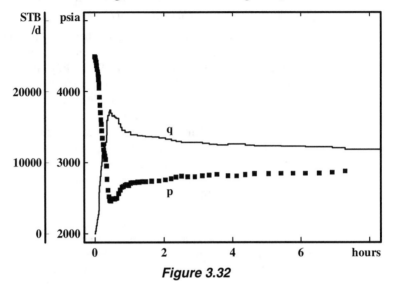

Figure 3.32

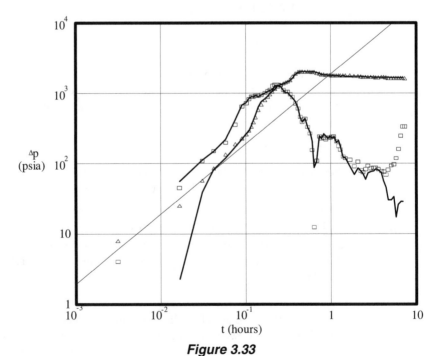

Figure 3.33

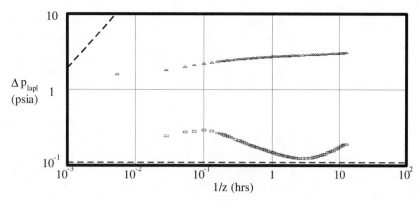

Figure 3.34

3.8 Interpretation Sequence

Having introduced the aspects of computer-aided interpretation separately, we can summarize the steps required in a typical test. These steps are illustrated pictorially in Fig. 3.35.

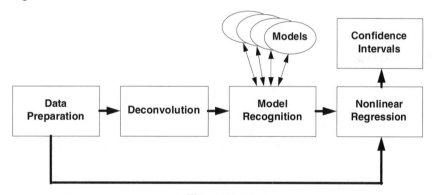

Figure 3.35

(a) **Data Preparation:** Pressure and flow rate data are sampled to reduce their number, and datum shifted to correct early time inaccuracies (Section 3.5)

(b) **Deconvolution:** The equivalent constant rate test response is extracted as an aid to diagnosis (Section 3.7).

(c) **Diagnosis and Model Identification:** An appropriate reservoir model (or a number of possible candidates) is selected based upon identifiable characteristics (Section 3.3). A preliminary estimate of reservoir parameter values may be made to initiate nonlinear regression.

(d) **Nonlinear Regression:** A best fit to the chosen model is obtained, and confidence intervals estimated (Section 3.6). Note that it is the *original* data that are used for nonlinear regression, not the deconvolved data (which have had the useful flow rate information extracted).

(e) **Confidence Interval Evaluation:** After reaching the best fit available for a given model, the confidence intervals should be evaluated to determine whether the interpretation is valid or not (Section 3.6.1). If more than one candidate reservoir model has been tried, confidence interval analysis may be useful to discriminate between candidate models. It should be noted that confidence interval values themselves are not comparable from one model to another -- rather than comparing values, the confidence intervals may sometimes be used to identify one model that is acceptable while others are not, even though all of the models may match the data.

It is worth noting that the overall procedure is general, since it is the same for all kinds of test: drawdown, buildup, injection, falloff, multirate and drillstem test (DST). The fact that nonlinear regression works as well with multirate tests as it does with constant rate tests avoids many of the specialized analyses made necessary by the restrictions required by traditional graphical techniques. Computer-aided interpretation is not only faster, more accurate and more widely applicable, it is also a good deal simpler.

3.9 Non-Ideality

The diffusive nature of pressure transmission in porous media means that small scale effects are hidden, and broad scale averaged properties most strongly influence a pressure transient response. Therefore a surprisingly large percentage of actual well tests really do respond in the manner suggested by the reservoir models (perhaps 70 to 80%). Nonetheless, there are physical and mechanical difficulties that can cause the pressure response to differ from that predicted by any reservoir model. These excursions can be due to gauge failure, wellbore transients (such as phase redistribution), unregistered flow changes, neglected reservoir mechanisms (such as segregated flow), human error during recording, etc. In some cases these non-idealities can lead to a response that is atypical in its entirety but in other cases there may be parts of the response that appear normal, and other parts that are strange (Fig. 3.36).

Such regions of abnormal response are of lesser importance in traditional graphical methods, which use only specific portions of the data. However, nonlinear regression techniques, which match the whole data set, are sometimes adversely affected. In attempting to honor the abnormal data, a regression may be forced to misfit the normal part of the data (Fig. 3.37).

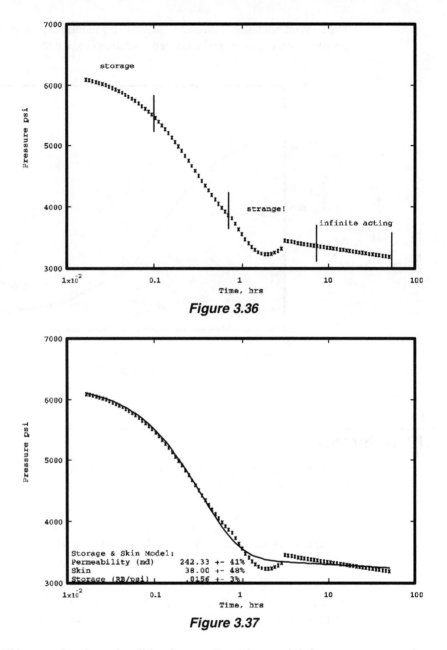

Figure 3.36

Storage & Skin Model:
Permeability (md)   242.33 +- 41%
Skin   38.00 +- 48%
Storage (RB/psi)   .0156 +- 3%

Figure 3.37

This may give rise to invalid estimates. In such cases it is important to recognize and *remove* the abnormal parts of the data, after which an automated analysis can determine those parameters for which it has appropriate ranges of data (Fig. 3.38).

In the example illustrated here, the correct reservoir permeability of 87.67 md is only obtained when the abnormal data (probably due to phase redistribution) are removed.

Thus a valid nonlinear regression analysis may require the intervention of the interpretation engineer, and some repreparation of the data.

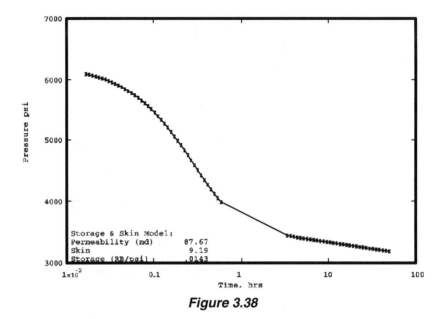

Figure 3.38

3.10 References

Bourdet, D., Whittle, T.M., Douglas, A.A., and Pirard, Y-M.: "A New Set of Type Curves Simplifies Well Test Analysis", *World Oil*, (May 1983), 95-106.

Bourdet, D., Ayoub, J.A., Whittle, T.M., Pirard, Y-M., and Kniazeff, V.: "Interpreting Well Tests in Fractured Reservoirs", *World Oil*, (October 1983), 77-87.

Bourdet, D., Ayoub, J.A., and Pirard, Y-M.: "Use of the Pressure Derivative in Well Test Interpretation", *SPE Formation Evaluation*, (June 1989), 293-302.

Bourgeois, M., and Horne, R.N.: "Model Identification Using Laplace Space Type Curves", *SPE Formation Evaluation*, (March 1993), 17-25.

Economides, C.A, Joseph, J.A., Ambrose, R.W. and Norwood, C,: "A Modern Approach to Well Test Interpretation", paper 19814, presented at the 64th SPE Annual Technical Conference and Exhibition, San Antonio, TX, Oct. 8-11, (1989), 429-444.

Gladfelter, R.E., Tracy, G.W., and Wilsey, L.E.: "Selecting Wells Which Will Respond to Production-Stimulation Treatment", *Drill. and Prod. Prac.*, API (1955), 117-129.

Gringarten, A.C.: "Computer-Aided Well Test Analysis", paper SPE 14099 presented at the SPE International Meeting on Petroleum Engineering, Beijing, China, March 17-20, (1986).

Guillot, A., and Horne, R.N.: Using Simultaneous Downhole Rate and Pressure Measurements to Improve Analysis of Well Tests", SPE Formation Evaluation, (1986), 217-226.

Horne, R.N., and Kuchuk, F.: "The Use of Simultaneous Flow Rate and Pressure Measurements to Replace Isochronal Gas Well Tests", *SPE Formation Evaluation*, (June 1988), 467-470.

Horne, R.N.: "Advances in Computer-Aided Well Test Interpretation", *J. Petroleum Tech.*, (July 1994), 599-606.

Kuchuk, F., and Ayestaran, L.: "Analysis of Simultaneously Measured Pressure and Sandface Flow Rate in Transient Well Testing", *J. Pet. Tech.*, (Feb. 1985), 323-334.

Kuchuk, F.: "Applications of Convolution and Deconvolution to Transient Well Tests", *SPE Formation Evaluation*, (Dec. 1990), 375-382.

Mendes, L.C.C., Tygel, M., and Correa, A.C.F.: "A Deconvolution Algorithm for Analysis of Variable Rate Well Test Pressure Data", paper 19815, presented at the 64th SPE Annual Technical Conference and Exhibition, San Antonio, TX, Oct. 8-11, (1989), 445-459.

Raghavan, R.: "*Well Test Analysis*", Prentice Hall, NJ (1993).

Roumboutsos, A., and Stewart, G.: "A Direct Deconvolution or Convolution Algorithm for Well Test Analysis", paper 18157, presented at the 63rd SPE Annual Technical Conference and Exhibition, Houston, TX, Oct. 2-5, (1988).

Winestock, A.G., and Colpitts, G.P.: "Advances in Estimating Gas Well Deliverability", *Jour. Cdn. Pet. Tech.*, (July-Sept. 1965), 111-119.

4. GAS WELL TESTS

4.1 Introduction

The analysis of gas well tests is made more complex by the fact that gas properties are strong functions of pressure, hence the equations governing pressure transmission through gases in a porous medium are nonlinear. Since all of the solutions derived for pressure transient analysis of liquid filled reservoirs are based on the slightly compressible pressure transmission equation (Eq. 2.2), it may seem at first as if none of these solutions would be applicable to the interpretation of gas well tests. Fortunately, by suitable definition of alternative variables, specifically the use of pseudopressure and pseudotime instead of pressure and time, most of the slightly compressible solutions can be modified for application to gas well test analysis.

4.2 Real Gas Pseudopressure and Pseudotime

The viscosity μ and compressibility c_g of real gases are strong functions of pressure, and it is not correct to apply the normal (slightly compressible) assumption when deriving the differential equations governing the pressure transients. However, if the gas is treated as obeying the real gas equation:

$$pV = znRT \tag{4.1}$$

then the governing differential equations can be linearized (approximately) by the definition of a variable termed the **real gas pseudopressure** by Al-Hussainy and Ramey (1966), and Al-Hussainy, Ramey and Crawford (1966). The real gas pseudopressure is defined as:

$$m(p) = 2 \int_{p_0}^{p} \frac{p}{\mu z} dp \tag{4.2}$$

where the base pressure p_0 is an arbitrary pressure, usually at the lowest end of the range of pressures of interest during the test.

This definition of the real gas pseudopressure results in an equation governing pressure transmission as:

$$\frac{\partial^2 m(p)}{\partial r^2} + \frac{1}{r}\frac{\partial m(p)}{\partial r} = \frac{\phi\mu c_t}{k}\frac{\partial m(p)}{\partial t}$$

(4.3)

This equation is linear with respect $m(p)$, except for the fact that the term $\phi\mu c_t/k$ is still a function of pressure (and therefore of pseudopressure). In practice, this remaining nonlinearity is not usually of consequence, and is often permissible to treat the equation as linear, substituting the values of viscosity μ and compressibility c_t defined at the initial reservoir pressure p_i (or at the highest pressure measured during the test if the initial reservoir pressure is not known). In cases where the gas compressibility variations with pressure are significant (such as may occur if the gas pressures are very low e.g. < 100 psia), the equations can be linearized further by the introduction of the **pseudotime**.

Pseudotime was introduced by Agarwal (1979) and its use described by Lee and Holditch (1982). Agarwal's pseudotime is defined as:

$$t_{pseudo} = \int_0^t \frac{1}{\mu c_t}\,dt$$

(4.4)

The definition of pseudotime is unintuitive to use, since it does not have units of time and thus makes visual examination of the data plots more difficult.

It may noted in Eqs. 4.2 and 4.4 that if μz is approximately constant then the pseudopressure becomes like the square of pressure, and the pseudotime becomes like the time itself. This is the origin of earlier gas well test analysis techniques that made use of **pressure-squared** plots. Pressure-squared analysis has the advantage of simplicity for hand calculation as compared to pseudofunction analysis, however since most modern analysis techniques are already computer-based there is little to be gained by not using the more accurate pseudofunction approach. The pressure-squared analysis equations will not be discussed here, they may found in Chapter 5 of the book by Sabet (1991) if required.

Both pseudopressure and pseudotime are somewhat awkward variables to work with, since they have large magnitude and strange units. The large magnitude of the pseudopressure can sometimes cause difficulty in computation, since pressure drops may be represented as small differences between large numbers. A new formulation for pseudopressure and pseudotime was introduced by Meunier, Kabir and Wittman (1987). These new normalized pseudofunctions allow the use of liquid flow solutions without special modification for gas flow, since the newer definitions have the same units as pressure and time. The Meunier *et al* definitions of **normalized pseudopressure** and **normalized pseudotime** are:

$$p_{pn} = p_i + \frac{\mu_i z_i}{p_i}\int_{p_i}^{p}\frac{p}{\mu z}\,dp$$

(4.5)

$$t_{pn} = \mu_i c_{ti} \int_0^t \frac{1}{\mu c_t} \, dt$$

(4.6)

In Eqs. 4.5 and 4.6 the subscript i on μ, z and c_t refers to the evaluation of these parameters at the initial pressure p_i. By using the Meunier *et al* definitions of normalized pseudopressure and normalized pseudotime there is no need to modify any of the liquid analysis equations. It is often not necessary to use pseudotime at all, so most interpretations can be analyzed more conveniently in terms of real time -- in cases of low gas pressures, the Meunier *et al* normalized pseudotime can be used as a direct replacement of the time variable.

Other than the very real advantage that liquid well test analysis equations can all be used without modification, an additional attraction of the normalized pseudopressure is that it has units of pressure and has a magnitude much like that of the original pressure. At the value p_i, the pressure and the normalized pseudopressure are the same. Similarly, the pseudotime has the same units as time itself. From the point of view of the interpretation engineer, this means that normalized pseudopressure plots look entirely conventional and the analysis is therefore easier to perform.

Since the use of real gas pseudopressures (the original definition, given by Eq. 4.2) is traditional, their application will be described here for completeness (see "4.5 Using Pseudopressure for Analysis" on page 113). The more convenient analysis, using normalized pseudopressure, will be described in "4.4 Using Normalized Pseudopressure for Anslysis" on page 112.

4.2 Calculating Pseudopressures

To calculate pseudopressures or normalized pseudopressures from a set of pressure data, it is necessary to evaluate an integral such as the one expressed by Eq. 4.2 or Eq. 4.5. Usually this is done numerically, since the properties may not be definable functions of pressure. The numerical integration is best done using a computer, and can be handled either by specially written programs, or by a spreadsheet approach. The steps involved are summarized in Table 4.1, which shows a spreadsheet calculation of the pseudopressure integral using input values of pressure p, viscosity μ, and z factor.

The integration shown in the spreadsheet uses the trapezoidal rule to evaluate:

$$m(p) = 2 \sum_{i=2}^{n} \frac{1}{2} \left[\left(\frac{p}{\mu z} \right)_{i-1} + \left(\frac{p}{\mu z} \right)_i \right] (p_i - p_{i-1})$$

(4.7)

p (psia)	visc (cp)	z	dp	p/(uz)	p/(uz) AVE	2.dp.AVE	m(p)
14.7	0.01198	0.9987		1.22864e3			
200	0.01235	0.9839	185	1.64593e4	8.84398e3	3.27758e6	3.27758e6
400	0.01277	0.9686	200	3.23389e4	2.43991e4	9.75964e6	1.30372e7
600	0.01319	0.9544	200	4.76624e4	4.00006e4	1.60003e7	2.90375e7
800	0.01362	0.9414	200	6.23934e4	5.50279e4	2.20112e7	5.10486e7
1000	0.01405	0.9296	200	7.65645e4	6.94790e4	2.77916e7	7.88402e7
1200	0.01451	0.9194	200	8.99517e4	8.32581e4	3.33032e7	1.12143e8
1400	0.01496	0.9107	200	1.02759e5	9.63555e4	3.85422e7	1.50686e8
1600	0.01542	0.9038	200	1.14806e5	1.08782e5	4.35130e7	1.94199e8
1800	0.01589	0.8986	200	1.26061e5	1.20434e5	4.81734e7	2.42372e8
2000	0.01636	0.8953	200	1.36546e5	1.31304e5	5.25214e7	2.94893e8

Table 4.1: Real gas pseudopressure calculation

Figure 4.1 illustrates the nonlinearity of pseudopressure as a function of pressure, by plotting the data from Table 4.1.

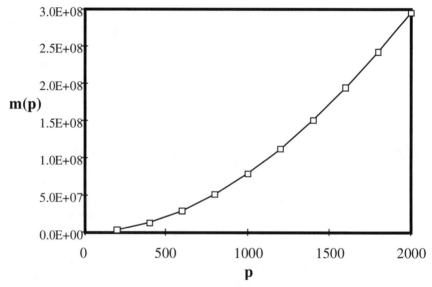

Figure 4.1 Plot of pseudopressure against pressure. Data from Table 4.1.

A similar spreadsheet shown in Table 4.2 shows the computation of normalized pseudopressure for the same set of pressure data. Figure 4.2 illustrates the nonlinearity of normalized pseudopressure as a function of pressure, by plotting the data from Table 4.2.

p (psia)	visc (cp)	z	dp	p/(uz)	p/(uz) AVE	integral	$p_{pn}(p)$
14.7	0.01198	0.9987	-185	1.22864e3	8.84398e3	-1.47447e8	920
200	0.01235	0.9839	-200	1.64593e4	2.43991e4	-1.45808e8	932
400	0.01277	0.9686	-200	3.23389e4	4.00006e4	-1.40928e8	968
600	0.01319	0.9544	-200	4.76624e4	5.50279e4	-1.32928e8	1026
800	0.01362	0.9414	-200	6.23934e4	6.94790e4	-1.21922e8	1107
1000	0.01405	0.9296	-200	7.65645e4	8.32581e4	-1.08027e8	1209
1200	0.01451	0.9194	-200	8.99517e4	9.63555e4	-9.13750e7	1331
1400	0.01496	0.9107	-200	1.02759e5	1.08782e5	-7.21039e7	1472
1600	0.01542	0.9038	-200	1.14806e5	1.20434e5	-5.03474e7	1631
1800	0.01589	0.8986	-200	1.26061e5	1.31304e5	-2.62607e7	1808
2000	0.01636	0.8953		1.36546e5			2000

Table 4.2 Normalized pseudopressure calculation

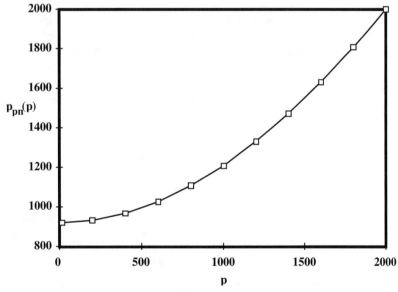

Figure 4.2 Plot of normalized pseudopressure against pressure. Data from Table 4.2

4.4 Using Normalized Pseudopressure for Analysis

Modern well test analysis permits a choice of variable for interpretation of gas well tests: pressure-squared, pseudopressure, or normalized pseudopressure. The pressure-squared approach is less commonly used in modern analysis, since it oversimplifies the dependence of gas properties on pressure. The pseudopressure approach has been used since the early 1970's, but has more recently been superseded by the normalized pseudopressure approach which is more convenient for the interpretation engineer as well as for the development of interpretation software.

Normalized pseudopressure can be used with all of the traditional graphical techniques described in Section 2 and with all of the computer-aided techniques described in Section 3. Nonetheless some care is required to make sure the appropriate scaling is used to specify the flow rates.

Many of the analysis equations include the flow rate term qB. Examples include Eqs. 2.3, 2.32, 2.28, 2.29, 2.38, 2.39, 2.68 and 2.69. It should be noted that the flow rate q always appears as a product with formation volume factor B since the term qB represents the volume of fluid actually removed from the reservoir. The analysis equations are generally written in terms of liquid volume removal, so it is essential that the formation volume factor B be defined appropriately. For example, consider an equation such as Eq. 2.3 in which the product qB is intended to represent the number of reservoir barrels per day (RB/d) removed from the reservoir -- if the gas flow rate q is measured in MCF/d then the formation volume factor B must be expressed in RB/MCF. This would therefore be a different unit than those commonly used for gas formation volume factor.

Once appropriate care has been taken with the expression of formation volume factor B, the conventional analysis equations can be used without modification.

One additional note of caution should be heeded when using normalized pseudopressures. The normalized pseudopressure is dependent on the value of the initial pressure p_i used to evaluate the integral definition, however the initial reservoir pressure may be one of the parameters estimated during the interpretation (see "3.6.2 Initial Pressure" on page 94). It is important not to "recycle" the value of p_i by recomputing the normalized pseudopressures, otherwise parameter estimates (particularly skin and initial reservoir pressure) are likely to drift continuously. It is not a requirement that the value of p_i used to compute the normalized pseudopressures be exactly the initial reservoir pressure -- the value is simply a pressure point at which the pressure and normalized pseudopressure coincide. Once an estimate of the "initial" value of normalized pseudopressure has been found from the analysis, this value can be converted to initial reservoir pressure by use of a conversion table such as Table 4.2 or conversion chart such as Fig. 4.2.

4.5 Using Pseudopressure for Analysis

The pseudopressure transient equation (Eq. 4.3) is exactly the same in form as the pressure transient equation (Eq. 2.2) used for oil and water well transient analysis. This means that all of the liquid (slightly compressible) solutions can also be used for the analysis of gas well tests, provided the equations are modified to accommodate the differences between pseudopressure and pressure.

In the same manner as the dimensionless pressure was defined to nondimensionalize the differential equation, a dimensionless pseudopressure can be defined for use with gas wells:

$$m_D = \frac{1.987 \times 10^{-5} \, kh \, T_{sc} [m(p_i) - m(p)]}{p_{sc} T \, q_{sc}}$$

(4.8)

where p_{sc} and T_{sc} are the pressure and temperature at standard conditions (usually 14.7 psia and 60 °F or 520 °R), and q_{sc} is the gas flow rate measure at standard conditions, measured in MCF/d. It is important to note that both the temperatures in this equation, T and T_{sc}, are expressed in absolute units (°R), since they originate from the real gas equation.

The dimensionless time is defined the same as for liquid wells, however it should be remembered that viscosity μ and compressibility c_t are to be evaluated at initial reservoir pressure.

$$t_D = \frac{0.000264 \, kt}{\phi (\mu c_t)_i r_w^2}$$

(4.9)

Once dimensionless variables are defined in this way, then m_D can replace p_D in any of the solutions or type curves developed for slightly compressible fluid transients.

In semilog analysis (MDH or Horner) the slope of the semilog straight line during infinite acting behavior, when the flow rate q_{sc} is in MCF/d is given as:

$$m = \frac{5.794 \times 10^4 \, q_{sc} p_{sc} T}{kh T_{sc}}$$

(4.10)

(Be careful not to be confused by the use of the symbol m for the slope; this is not the pseudopressure, which is always written $m(p)$). Eq. 4.10 is used as before, to estimate permeability-thickness kh. The skin factor can be calculated from:

$$s = 1.151 \left[\frac{m(p_i) - m(p_{1hr})}{m} - \log \frac{k}{\phi(\mu c_t)_i r_w^2} + 3.2274 \right]$$

(4.11)

or for a Horner plot:

$$s = 1.151 \left[\frac{m(p_{1hr}) - m(p_{wf})}{m} - \log \frac{kt_p}{(t_p + 1)\phi(\mu c_t)_i r_w^2} + 3.2274 \right]$$

(4.12)

Wellbore storage coefficient C can be evaluated from a point $[\Delta t, \Delta m(p)]$ on the unit slope, log-log line as:

$$C = \frac{83.39 p_{sc} T q_{sc} \Delta t}{T_{sc} \mu_i \Delta m(p)}$$

(4.13)

The dimensionless storage coefficient C_D is calculated from the estimate of C as before:

$$C_D = \frac{C}{2\pi\phi(c_t)_i h r_w^2}$$

(4.14)

Type curve matching is performed just as in the liquid well case, using the definitions of the dimensionless variables, Eqs. 4.8 and 4.9, to estimate kh. For example, if $m(p)$ is a pseudopressure data point, and m_D is the corresponding point on the type curve after matching, then:

$$kh = \frac{p_{sc} T q_{sc} m_D}{1.987 \times 10^{-5} T_{sc} m(p)}$$

(4.15)

For boundary analysis, m_D is used exactly as was p_D for liquid well cases. So, for example, during pseudosteady state the drainage area can be estimated by using:

$$m_D = 2\pi t_{DA} + \frac{1}{2} \ln \left[\frac{2.2458A}{C_A r_w^2} \right] + s$$

(4.16)

4.6 Rate Dependent Skin Effect

A remaining difficulty in gas well test interpretations lies in the nonlinearity of the boundary conditions, and this is not easily overcome. The difficulty arises due to the "turbulent" or non-Darcy flow effects close to the wellbore, that appear as a rate-dependent skin as described by Muskat (1946), by Katz et al (1959) and by Wattenbarger and Ramey (1968). The total skin effect is thus comprised of a constant part s' together with a rate dependent part Dq_{sc}:

$$s = s' + Dq_{sc}$$

(4.17)

The constant part of the skin factor s' is due to the wellbore damage effect, while the rate dependent part is a function of the flow rate. To correctly identify the condition of the well, as well as to evaluate the absolute open flow potential (AOFP), it is necessary to be able to evaluate the two different skin effects separately. It is this estimation of the rate dependent effect that requires gas wells to be tested at a variety of rates, using the well known flow-after-flow test, Rawlins and Schellhardt (1936), isochronal test, Cullender (1955) or modified isochronal test, Katz et al (1959), Aziz (1967), and Brar and Aziz (1978). Of these methods, only the modification of Brar and Aziz (1978) avoids the necessity of running the test until flow stabilization is achieved. Thus all of these methods of interpretation require well tests of quite long duration. Recently, Meunier, Kabir and Wittmann (1987) showed how the inclusion of flow data into the analysis allows the test to be of much shorter duration. This important study demonstrated the advantages of including additional measurements into pressure transient tests as proposed earlier by Kuchuk and Ayestaran (1985). Meunier, Kabir and Wittmann (1987) used a "rate-normalized quasipressure" to reduce the variable rate test to an equivalent single rate test. This idea permits such a test to be analyzed as if the test had been at constant rate. Unfortunately, the non-Darcy skin effect must be included as a correction in the Meunier, Kabir and Wittmann (1987) "rate-normalized quasipressure", after which it is necessary to substitute values of the non-Darcy skin parameter by trial and error until a semilog straight line is obtained. There are two disadvantages in doing this: firstly the reservoir may show a response other than radial flow (dual porosity or fracture flow would be common alternatives in tight gas reservoirs) and secondly the trial and error procedure is awkward to perform.

Nonlinear regression provides a more straightforward method of estimating the rate dependent skin factor D, provided the well has been flowed at different rates during the test. In cases where a downhole flow measurement has been made, it has even been possible to estimate D and $AOFP$ with a single rate test, as shown by Horne and Kuchuk (1988).

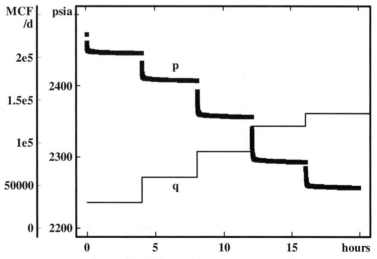

Figure 4.3 Example flow-after-flow gas well test.

As an example of the effect of non-Darcy skin, consider the flow-after-flow test shown in Fig. 4.3. Four of the individual flow periods are plotted using a rate-normalized plot in Fig. 4.4, revealing that the skin factor is increasing with increasing flow rate (if the skin factors were the same, the semilog straight lines in each of the flow periods would be aligned with each other). Examining the skin factor as a function of flow rate in Fig. 4.5 shows that the $s'+Dq$ behavior suggested from Eq. 4.17 is seen. The damage skin s' can be found from the intercept of the line, and the non-Darcy skin D can be found from the slope.

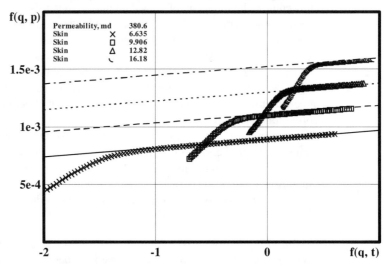

Figure 4.4 Rate normalization plot of four of the flow periods from Fig. 4.3.

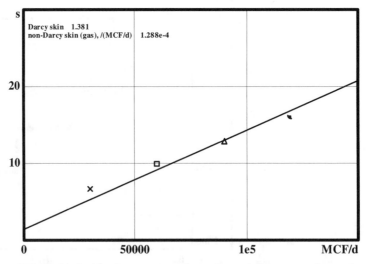

Figure 4.5 Plot of skin factor as a function of flow rate, based on the four skin factor estimates from Fig. 4.4.

4.7 References

Agarwal, R. G.: "Real Gas Pseudotime - A New Function for Pressure Buildup Analysis of Gas Wells", paper SPE 8279 presented at the 1979 SPE Annual Technical Conference and Exhibition, Las Vegas, Nevada, September 23-26, 1979.

Al-Hussainy, R. and Ramey, H. J. Jr.: "Application of Real Gas Flow Theory to Well Testing and Deliverability Forecasting", *J. Pet. Tech.* (May 1966) 637-642; Trans., AIME, **237**.

Al-Hussainy, R., Ramey, H. J. Jr., and Crawford, P.B.: "The Flow of Real Gases Through Porous Media", *J. Pet. Tech.* (May 1966), 624-636; Trans. AIME, 237.

Aziz, K.: "Theoretical Basis of Isochronal and Modified Isochronal Back-Pressure Testing of Gas Wells", *J. Can. Pet. Tech.* (January-March 1967) 20-22.

Brar, G. S. and Aziz, K.: "Analysis of Modified Isochronal Tests to Predict the Stabilized Deliverability Potential of Gas Wells Without Using Stabilized Flow Data", *J. Pet. Tech.* (February 1978) 297-304.

Cullender, M. H.: "The Isochronal Performance Method of Determining the Flow Characteristics of Gas Wells", *Trans.*, AIME **204**, (1955), 137-142.

Horne, R.N., and Kuchuk, F.: "The Use of Simultaneous Flow Rate and Pressure Measurements to Replace Isochronal Gas Well Tests", *SPE Formation Evaluation*, (1988), 467-470.

Katz, D. L., Cornell, D., Kobayashi, R., Poettmann, F. H., Vary, J. A., Elenbaas, J. R. and Weinaug, C. F.: *Handbook of Natural Gas Engineering*, McGraw-Hill Book Co. Inc., New York, 1959.

Kuchuk, F. and Ayestaran, L.: "Analysis of Simultaneously Measured Pressure and Sandface Flow Rate in Transient Well Testing", *J. Pet. Tech.* (February 1985) 322-330.

Lee, W. J. and Holditch, S. A.: "Application of Pseudotime to Buildup Test Analysis of Low Permeability Gas Wells with Long-Duration Wellbore Storage Distortion", J. Pet. Tech. (December 1982) 2877-2887.

Meunier, D., Kabir, C. S. and Wittmann, M. J.: "Gas Well Test Analysis: Use of Normalized Pressure and Time Functions", *SPE Formation Evaluation*, (December 1987), 629-636.

Muskat, M.: *The Flow of Homogeneous Fluids*, J. E. McEdwards, Inc., Ann Arbor, Mich. (1946).

Rawlins. E. L. and Schellhardt, M. A.: "Backpressure Data on Natural Gas Wells and Their Application to Production Practices", U. S. Bureau of Mines, Monograph, 7, (1936).

Sabet, M.: *Well Test Analysis*, Gulf Publishing, Houston, (1991).

Wattenbarger, R. A. and Ramey, H. J., Jr.: "Gas Well Testing with Turbulence, Damage and Wellbore Storage", *J. Pet. Tech.* (August 1968) 877-887.

5. MULTIPHASE WELL TESTS

5.1 Introduction

All of the preceding sections have examined the interpretation of tests of wells that produce single phase fluid - either oil, water or gas. In general, when more than one fluid phase flows in the reservoir at the same time, the multiphase interactions render the single phase flow equations invalid, and it may be necessary to develop new equations to specifically include the multiphase effects. Examples of situations in which this may arise are tests in solution gas drive reservoirs with well flowing pressures below the bubble point, or gas condensate reservoirs with well flowing pressures in the retrograde condensation region. When two or more phases flow simultaneously, the presence of one phase reduces the flow of the other, due to relative permeability effects. The result is that the effective permeabilities are functions of saturation, and therefore also of time. Since the total mobility is a combination of the separate phase mobilities, it follows that the net effective reservoir mobility will change during a well test and after it has been completed. To obtain useful interpretive results, these effects must be considered.

This section discusses different ways in which multiphase flow effects can be accommodated during a well test interpretation, however it is important to note at the outset that most existing techniques have restrictions that limit their applicability or accuracy. Therefore it is almost always better to design a well test ahead of time to avoid reaching multiphase conditions during the test. The most problematic of all cases is when the reservoir pressure passes through the bubble point pressure during the test -- this represents a very major change in flow conditions and is very hard to account for properly in the interpretation. If a well must be tested below the bubble point pressure, it should if at all possible be held below the bubble point pressure throughout the entire test.

Several different approaches to analysis of multiphase well tests have been developed. The traditional approach is due to Perrine (1956) and Martin (1959). Their approach makes use of the concepts of total mobility and total compressibility, and is still widely used due to its straightforward application. However, Perrine's approach is known to be less reliable as gas saturations increase, as shown by Weller (1966), and can underestimate the effective phase permeabilities as shown by Chu, Reynolds and Raghavan (1986). Ayan and Lee (1988) also found that

Perrine's approach overestimated the skin effect in cases where flow was blocked by gas in the vicinity of the wellbore.

A second approach to multiphase well test analysis by Raghavan (1976) makes use of specially defined pseudopressure that is analogous to the real gas pseudopressure derived for gas wells in "4.2 Real Gas Pseudopressure and Pseudotime" on page 107. An overview of this approach in a broader context was also described by Raghavan (1986). The definition of multiphase pseudopressure is:

$$m(p) = \int_0^p \frac{k_{ro}}{\mu_o B_o} \, dp$$

(5.1)

This definition was used by Fetkovitch (1973) to define the productivity of wells:

$$q_o = \frac{kh}{141.2(0.5 \ln t_D + 0.404 + s)} [m(p_i) - m(p_{wf})]$$

(5.2)

where:

$$t_D = \frac{0.000264kt}{\phi \mu_o c_t r_w^2}$$

(5.3)

Application of the pseudopressure approach requires knowledge of the relative permeability curves appropriate for the reservoir. Aanonsen (1985a) and Aanonsen (1985b) demonstrated that small inaccuracies in the relative permeability data can lead to greater inaccuracies in reservoir parameter estimates. This is a weakness of the pseudopressure approach, since in practice reservoir relative permeability data is quite difficult to obtain.

A third approach was described by Al-Khalifah, Aziz and Horne (1987), who developed a procedure that generalized the works of Perrine (1956), Martin (1959) and Fetkovitch (1973) by developing them from first principles. This approach is based on the use of pressure squared p^2 instead of pressure, and avoids the need to know the reservoir relative permeability behavior in advance. It has been successfully demonstrated for both high volatility and low volatility oil systems.

In this section, we will review first the traditional Perrine approach, since it still in common usage, and then describe the more general pressure squared approach of Al-Khalifah, Aziz and Horne (1987).

5.2 Perrine's Approach

The essence of traditional analysis of multiphase tests is the definition of total flow properties to replace the individual phase properties. The interpretation procedure uses the **total mobility** λ_t, and the **total compressibility** c_t, defined as follows:

$$\lambda_t = \frac{k_o}{\mu_o} + \frac{k_w}{\mu_w} + \frac{k_g}{\mu_g}$$

(5.4)

$$c_t = c_r + S_o c_o + S_w c_w + S_g c_g + \frac{S_o B_g}{5.615 B_o}\left(\frac{\partial R_s}{\partial p}\right) + \frac{S_w B_g}{5.615 B_w}\left(\frac{\partial R_{sw}}{\partial p}\right)$$

(5.5)

where the phase permeability is defined in terms of the relative permeability:

$$k_o = k\, k_{ro}; \quad k_w = k\, k_{rw}; \quad k_g = k\, k_{rg}$$

(5.6)

In the definition of total compressibility c_t, the last two terms on the right hand side of Eq. 5.5 represent the volume change due to gas dissolution upon pressure change. In practice, these terms are large (although the water-gas solution term is small relative to the oil-gas term), and can be the major contributor to the compressibility. This means that multiphase systems undergoing gas evolution from oil can have very large compressibilities, often larger than systems having only gas.

In the case of gas condensate reservoirs, the total compressibility is modified to take account of the formation of condensate:

$$c_t = c_r + S_o c_o + S_w c_w + S_g c_g + \left(\frac{\partial r_s}{\partial p}\right) + \frac{S_w B_g}{5.615 B_w}\left(\frac{\partial R_{sw}}{\partial p}\right)$$

(5.7)

where r_s is the volume of condensate liquid at reservoir conditions relative to the total pore volume (this is usually the same as oil saturation S_o).

The definition of total mobility and compressibility does not fully linearize the governing equations, and their use is subject to some assumptions.

(a) Pressure gradients must be small.

(b) Saturation gradients must be small.

(c) Saturation changes during the duration of the test must be negligible.

(d) Capillary pressure must be negligible.

Provided these conditions are satisfied, the methods and solutions developed for single phase well test applications can be used for multiphase well test interpretation. It is necessary to make suitable substitutions in the equations used for the analysis, being certain to use the multiphase definition of total compressibility and replacing terms in k/μ by the total mobility λ_t. It is also necessary to use the total fluid production rate as follows:

$$(qB)_t = q_o B_o + q_w B_w + [1000 q_g - R_s q_o - R_{sw} q_w] \frac{B_g}{5.615} \qquad (5.8)$$

For example, when performing type curve analysis, the definitions of dimensionless pressure and time are:

$$p_D = \frac{\lambda_t h}{141.2 (qB)_t} \Delta p \qquad (5.9)$$

$$t_D = \frac{0.000264 \lambda_t t}{\phi c_t r_w^2} \qquad (5.10)$$

As another example, in semilog analysis, the slope of the infinite acting semilog straight line and the skin effect are given respectively by:

$$m = \frac{162.6 (qB)_t}{\lambda_t h} \qquad (5.11)$$

$$s = 1.151 \left[\frac{p_i - p_{1hr}}{m} - \log \frac{\lambda_t}{\phi c_t r_w^2} + 3.227 \right] \qquad (5.12)$$

As a result of these definitions, the analysis procedure provides an estimate of total mobility λ_t rather then permeability directly. If the relative permeabilities of the reservoir are known, then the separate phase mobilities can be estimated from the fractional flow curves, for example:

$$\frac{\lambda_w}{\lambda_o} = f_w (S_w) \qquad (5.13)$$

$$\frac{\lambda_g}{\lambda_o} = f_g (S_g) \qquad (5.14)$$

Alternatively, the calculation can use the phase relative permeabilities k_{ro}, k_{rw} and k_{rg} directly, making use of Eqs. 5.4 and 5.6. In either method, it is necessary to know the saturations of each of the phases in the reservoir. This may present something of a difficulty since the saturations originally measured by logging will be different from those under the conditions of the well test. Saturations may also vary during the test and will be different at different distances from the well. Bøe, Skjaeveland and Whitson (1981) developed an expression that relates saturation to pressure during drawdown in solution gas reservoirs. This expression may be used to estimate saturations for use in Eqs. 5.13 and 5.14.

Absolute reservoir permeability can also be estimated from λ_t by substituting the relative permeability values in the expanded form of Eq. 5.4:

$$\lambda_t = k\left[\frac{k_{ro}}{\mu_o} + \frac{k_{rw}}{\mu_w} + \frac{k_{rg}}{\mu_g}\right]$$

$$(5.15)$$

5.3 Pressure Squared Approach

As discussed by Al-Khalifah, Aziz and Horne (1987), the equations governing multiphase pressure transients during a well test can be reduced to the following equation in terms of pressure squared:

$$\frac{\partial^2 p^2}{\partial r^2} + \frac{1}{r}\frac{\partial p^2}{\partial r} = \frac{\phi c_t}{\lambda_t}\frac{\partial p^2}{\partial t}$$

$$(5.16)$$

This equation holds only provided the group of parameters $k_o/(\mu_o B_o)$ is a linear function of pressure:

$$\frac{k_o}{\mu_o B_o} = \alpha p$$

$$(5.17)$$

In practice this is often approximately true, as has been reported by Handy (1957), Fetkovitch (1973) and Al-Khalifah, Aziz and Horne (1987).

Using these expressions, a new set of equations can be developed for interpretation. For example, the equation describing the infinite acting (semilog straight line) behavior can be written:

$$p_i^2 - p_{wf}^2 = \frac{325.1 q_o}{\alpha h}\left[\log t + \log\left(\frac{\lambda_t}{\phi c_t r_w^2}\right) - 3.227 + 0.868s\right]$$

$$(5.18)$$

The skin factor can be found from:

$$s = 1.151\left[\frac{p_i^2 - p_{1hr}^2}{m} - \log\left(\frac{\lambda_t}{\phi c_t r_w^2}\right) + 3.227\right]$$

$$(5.19)$$

where m is the slope of the infinite acting semilog straight line on the plot of pressure squared p^2 against the log of time.

After finding the slope of the infinite acting semilog straight line, the parameter α can be estimated, and used with Eq. 5.17 to estimate the oil phase permeability k_o. The question remains as to which value of pressure should be associated with the estimated value of α. Al-Khalifah, Aziz and Horne (1987) showed that if the average pressure during the infinite acting period is used, then the permeability calculated is the same as that of Perrine's approach (which usually underestimates oil permeability). As an alternative, it was proposed that α be evaluated at a higher pressure, specifically at initial reservoir pressure p_i for drawdown tests, and at average reservoir pressure $\bar{p}$ for buildup tests. Thus for drawdown tests, α would be given by:

$$\alpha = \left(\frac{k_o}{\mu_o B_o}\right)_i \frac{1}{p_i}$$

(5.20)

and oil phase permeability would therefore be given by:

$$k_o = \frac{325.2 \, q_o p_i (\mu_o B_o)_i}{mh}$$

(5.21)

For buildup tests, the expression for oil phase permeability would be:

$$k_o = \frac{325.2 \, q_o \bar{p} (\overline{\mu_o B_o})_i}{mh}$$

(5.22)

Al-Khalifah, Aziz and Horne (1987) showed that Eqs. 5.21 and 5.22 gave good estimates of oil phase permeability for tests involving oils of both high and low volatility. However, in cases of large drawdowns in tests with oils of low volatility, it was found to be necessary to use a lower pressure for the evaluation of α, otherwise the oil permeability was somewhat overestimated.

After estimating oil phase permeability k_o, water and gas phase permeabilities can be estimated in the same manner discussed in Section 5.2 on page 121, using Eqs. 5.13 and 5.14. Absolute reservoir permeability k can be estimated as before, using Eq. 5.15. Again, relative permeability data and knowledge of the phase saturations are required.

5.4 References

Aanonsen, S.I.: "*Nonlinear Effects During Transient Fluid Flow in Reservoirs as Encountered in Well Test Analysis,*" Dr. Scient. dissertation, Univ. of Bergen, Norway, 1985a.

Aanonsen, S.I.: "Application of Pseudotime to Estimate Average Reservoir Pressure", paper SPE 14256 presented at the 60th Annual SPE Technical Conference and Exhibition, Las Vegas, NV, Sept. 22-25, 1985b.

Ayan, C.J., and Lee, W.J.: "Multiphase Pressure Buildup Analysis: Field Examples", paper SPE 17412 presented at the SPE California Regional Meeting, Long Beach, CA, March 23-25, 1988.

Al-Khalifah, A.A., Aziz, K., and Horne, R.N.: "A New Approach to Multiphase Well Test Analysis", paper SPE 16473 presented at the 62nd Annual SPE Technical Conference and Exhibition, Dallas, TX, Sept. 27-30, 1987.

Bøe, A., Skjaeveland, S.M., and Whitson, C.S.: "Two-Phase Pressure Transient Test Analysis", paper SPE 10224 presented at the 56th Annual SPE Technical Conference and Exhibition, San Antonio, TX, Oct. 5-7, 1981.

Chu, W.C., Reynolds, A.C., Jr., and Raghavan, R.: "Pressure Transient Analysis of Two-Phase Flow Problems", *SPE Formation Evaluation*, (April 1986), 151-161.

Fetkovich, M.J.: "The Isochronal Testing of Oil Wells" paper 4529, presented at the SPE 48th Annual Fall Meeting, Las Vegas, NV, Sept. 30 - Oct. 3, 1973; *SPE Reprint Series*, No. 14, 265-275.

Handy, L.L.: "Effect of Local High Gas Saturations on Productivity Indices", *Drill. and Prod. Prac.*, API (1957).

Martin, J.C.: "Simplified Equations of Flow in Gas Drive Reservoirs and the Theoretical Foundation of Multiphase Buildup Analysis", *Trans.*, AIME (1959), **216**, 309-311.

Perrine, R.L.: "Analysis of Pressure Buildup Curves", *Drill. and Prod. Prac.*, API (1956), 482-509.

Raghavan, R.: "Well Test Analysis: Wells Producing by Solution Gas Drive", *Soc. Petr. Eng. J.*, (Aug. 1976), 1966-208.

Raghavan, R.: "Well Test Analysis for Multiphase Flow", paper SPE 14098 presented at the SPE International Meeting on Petroleum Engineering, Beijing, China, March 17-20, 1986.

Weller, W.T.: "Reservoir Performance During Two-Phase Flow", *J. Pet. Tech.*, (Feb. 1966), 240-246.

6. DESIGNING WELL TESTS

6.1 Introduction

As described in Section 1, a well test is intended to meet specific reservoir analysis objectives. To meet the required objectives, a test must be properly designed. An improperly planned test is not only a fruitless expense, it also fails to provide the desired reservoir data. In some cases, it may not be possible to attain the required objectives at all, and in other cases special equipment may need to be ordered in advance and transported to the wellsite. For all these reasons, it is essential to carefully consider what the test is to achieve, and how it is to be performed to successfully reach those goals.

Effective well test design requires consideration of which operational variables affect the estimates of which reservoir variables. For the most part, the operational variables under the control of the engineer are the flow rate and the duration of the test. A choice must also be made as to the type of test to be carried out. This section is divided into three parts, consideration of the effects of flow rate and time, specification of the test duration, and appraisal of the flow rate.

6.2 Variable Dependency

Which variables depend on which? Understanding this concept is of assistance in planning the test. In designing the operation, there are two major considerations; (a) will the reservoir parameter to be estimated affect the well pressure in a significant enough way that the effect will be detectable with the tools available to measure it, and the tools available to analyze the response, and (b) will the test be of sufficient length for the response to be seen.

We have seen in earlier sections, mainly Section 2, that different parts of the pressure response begin and end at certain times. These are summarized again here (in the order they would appear in an optimum case):

End of wellbore storage effect (Eq. 2.23):

$$t_D = C_D(0.041 + 0.02s)$$

(6.1)

Start of semilog straight line, either single porosity or secondary part of dual porosity (from Eq. 2.24):

$$t_D = C_D(60 + 3.5s)$$

(6.2)

End of secondary porosity semilog straight line in a dual porosity reservoir:

$$t_D = \frac{\omega(1-\omega)}{6.6\lambda}$$

(6.3)

End of dual porosity transition:

$$t_D = \frac{1.2(1-\omega)}{\lambda}$$

(6.4)

End of infinite acting behavior (semilog straight line) -- depends on reservoir shape and boundary configuration (see "2.8.1 Closed Boundaries" on page 30), for *circular* bounded reservoirs:

$$t_{DA} = 0.1$$

(6.5)

Start of pseudosteady state -- also depends on reservoir shape and boundary configuration, for *circular* bounded reservoirs:

$$t_{DA} = 0.1$$

(6.6)

The important feature to recognize about these start and end times is that they can all be expressed in terms of dimensionless time t_D or t_{DA}. This means that they are affected by mobility k/μ, storativity $\phi c_t h$ and transmissivity kh, **but not by flow rate**. Thus for any given reservoir, the transition times will occur at a specific moment, regardless of the rate at which the well is flowing.

The pressure response may be estimated by looking at specific models, but for illustration, we can look at the infinite acting response (Eq. 2.28):

$$p_{wf} = p_i - 162.6\frac{qB\mu}{kh}\left(\log t + \log\frac{k}{\phi\mu c_t r_w^2} + 0.8686s - 3.2274\right)$$

(6.7)

From this equation, it is seen that the magnitude of the pressure drop depends on the group of variables $qB\mu/kh$. Thus the pressure drop is directly proportional to the flow rate (and will always be so for single phase reservoirs). Deciding on a flow rate change to induce for the purpose of the test is not really an issue, since it is usually best to produce as large a response as practical, to be as sure as possible to obtain a recognizable set of characteristic responses. However it is necessary to determine in advance whether the likely pressure changes will be sufficient to be able to interpret the test adequately.

Nonetheless, we can see from Eq. 6.7 that the amount of the pressure change is only indirectly dependent on the storativity of the reservoir $\phi c_t h$. This means that reservoirs with high storativity will experience the same pressure drop as reservoirs with low storativity, although the times at which the pressure drop is reached will be very different.

6.3 Test Duration

The overall period of time for which the test must be conducted has to be sufficient to be assured of reaching that part of the reservoir response that is of interest. For example, if the objective of the test is to evaluate the effectiveness of an acidization treatment in removing skin damage, then it is necessary to obtain sufficient data past the end of the storage transition to be assured of recognizing the correct semilog straight line. Specifically, if we need an entire log cycle of infinite acting response, then the test would have to run for a time of at least (using Eq. 6.2):

$$t_D \geq 10 \times C_D(60 + 3.5s) \tag{6.8}$$

If, on the other hand, it was required to estimate the drainage area of the well, then we would need to observe a sufficient duration of pseudosteady state response, again perhaps a log cycle. Assuming the drainage area to be circular, we could use Eq. 6.6 to specify:

$$t_{DA} \geq 1 \tag{6.9}$$

Another consideration when examining test times is whether one part of the pressure response will overlap another. Perhaps the required part of the response will be disguised by a less diagnostic effect. For example, it is conceivable that a well with a large storage effect (such as caused by falling liquid level), in a relatively small reservoir, could exhibit a pressure transient that went from storage response directly to pseudosteady state. It could do this (in a circular reservoir) if:

$$C_D(60 + 3.5s) \geq 0.1 \frac{A}{r_w^2} \tag{6.10}$$

In such a case, the well would be difficult to interpret for permeability k and skin factor s if the wellbore storage coefficient were greater than:

$$C_D \geq \frac{A}{r_w^2} \frac{0.1}{(60 + 3.5s)}$$

(6.11)

In real dimensions, the storage coefficient C in STB/psi would need to be less than (using Eq. 2.17):

$$C \leq \frac{2\pi\phi c_t hA}{5.615} \frac{0.1}{(60 + 3.5s)}$$

(6.12)

If the wellbore storage is measured or estimated to be larger than this, then it will be pointless to conduct the well test unless the storage effect can be reduced or overcome (for example by using a downhole shutoff or by measuring flow rates down hole).

A similar consideration can be made to determine whether storage effects will make it impossible to interpret dual porosity effects, as will occur if:

$$C_D(60 + 3.5s) \geq \frac{\omega(1 - \omega)}{6.6\lambda}$$

(6.13)

6.4 Flow Rate Considerations

As discussed in Section 6.2, the amount of the pressure change is directly proportional to the flow rate in all but a few special cases. Thus it is necessary to determine whether the maximum flow rate change attainable will provide sufficient pressure change over the part of the response most diagnostic of the reservoir parameters of interest. For example, a highly permeable reservoir may have only very small pressure change during the semilog straight line period of its response -- if this change is so small as to be adversely influenced by measurement noise, then it will be difficult to obtain good estimates of the permeability. "8.6 Interference Test Examples" on page 194 shows data from a real well test example where this was the case.

Although flow rate may be chosen to maximize the resulting pressure response, it is undesirable for the well to pass through the bubble point pressure during the test since the resulting phase transition would make the test difficult to interpret. Therefore it is necessary to estimate in advance the magnitude of the expected pressure drop, and to reduce the flow rate if the pressure drop would be large enough to pass the bubble point pressure.

Since the pressure change is also a function of the reservoir permeability and fluid viscosity, and since the level of measurement noise depends on type of instrument used, it is difficult to prescribe a set of equations to evaluate whether the available flow will be sufficient. The best approach, especially in a computer-aided interpretation environment, is to "simulate" the test with prospective values of the reservoir parameters, and then examine the simulated data to see whether it is able to provide valid estimates of the required parameters or not. Some caution is required however, since "perfect" data can often provide answers over restricted data ranges, whereas real data will not. Thus it is necessary to collect more data than may be suggested from a simulated set of data, especially since it is also necessary to include contingencies and uncertainties.

Finally, it should also be noted that a well test in a computer-aided system of data acquisition and processing may be interpreted at the same time the test is being carried out, either at the wellsite or by using remote data transmission. Thus it is feasible to "redesign" the test before terminating the measurements, thereby saving the need for a retest while still meeting the original goals of the reservoir analysis.

7. ADVANCED TOPICS

7.1 Horizontal Wells

One of the most significant technological advances in the petroleum industry over the past decade has been the extensive development of horizontal drilling techniques. Horizontal wells are now common in many different applications, so it has become important to be able to interpret well tests in horizontal wells.

Horizontal wells differ from vertical wells in a number of ways that are important to well test interpretation:

(a) The open interval into which fluid enters the wellbore is very long. In many cases the producing length may not be well known, unless production logging has been used to measure the rate at which fluid enters the hole at different locations.

(b) Vertical permeability may play an important role, since there is likely to be considerable flow in the vertical direction.

(c) There are a number of different flow regimes during the transient, however depending on the values of the reservoir parameters one or more of the flow regimes may be missing.

Kuchuk (1995) presented an overview of interpretation methods for horizontal well tests, and his description will be summarized here.

7.1.1 Flow Regimes

Horizontal wells can exhibit a number of different flow regimes during their transient behavior. Depending on the magnitude of the reservoir parameters, one or more of the flow regimes may be missing. The parameters of significance to the nature of a horizontal well transient are the ratio of the vertical to horizontal permeability (k_V/k_H), the position of the well relative to the thickness of the formation (z_w/h), and the effective length of the well relative to the formation height (L_w/h).

Unless covered by wellbore storage effect, the first flow regime to be seen is the first radial flow, as illustrated in Fig. 7.1. During this period, the presence of the upper and lower boundaries have not yet been felt at the wellbore, so the well acts as it would in an infinite medium and the axis of the flow is radial (coincident with the axis of the wellbore). Although this is referred to as a radial flow period, it is important to acknowledge that the commonly large difference between vertical and horizontal permeability usually causes the flow to be elliptical rather than circular. If the formation height is small, or if k_V/k_H is small, this early radial flow may not be seen.

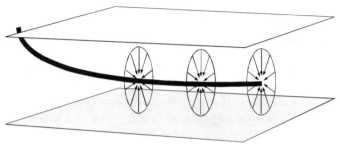

Early time radial flow

Figure 7.1

At late time, the influence of the upper and lower boundaries may cause another radial flow regime in which the axis of the flow is vertical, as illustrated in Fig. 7.2. Flow is radial in the horizontal plane, with the horizontal well acting somewhat like a point source. This occurs when the radius of investigation is large relative to the length of the horizontal well. The late time radial flow may not occur if other external boundaries are felt first, and will definitely not occur if the pressure is being supported by an aquifer or gas cap.

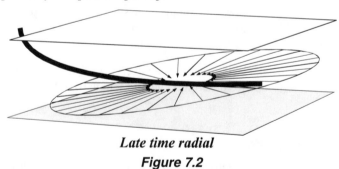

Late time radial

Figure 7.2

Between the early and late time radial flow periods there may be a linear flow regime, as illustrated in Fig. 7.3. In this regime the top and bottom boundaries exert their influence while the horizontal well length is still important relative to the radius of investigation. Linear flow may not occur if the formation is thick, or if k_V/k_H is small. Subsequent to early time radial flow, there may sometimes be an intermediate period of radial flow if the well is located close to one of the upper or lower boundaries. This is known as the hemiradial flow regime, and is illustrated in Fig. 7.4. This flow regime will usually not occur unless z_w/h is close either to zero or to one.

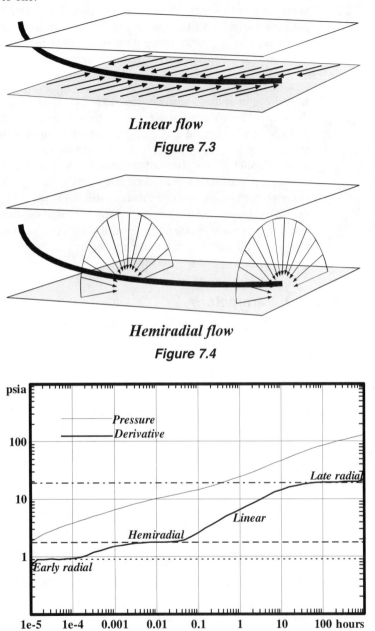

Linear flow

Figure 7.3

Hemiradial flow

Figure 7.4

Figure 7.5

An example transient demonstrating each of these flow periods is shown in Fig. 7.5, for $k_V/k_H = 0.05$, $z_w/h = 0.05$ and $L_w/h = 10$ [it should be noted here that the variable L_w is used here to represent the *total* effective well length -- this is different from the terminology used by Kuchuk (1995) in which L_w represents the well **half** length]. The transient in Fig. 7.5 covers a very wide range of time, and it would be unusual to see all four flow periods in a real well test. For example, in a real well test with reservoir parameters like those in Fig. 7.5, it is certain that the first radial flow will be too early to be visible during the test, and the hemiradial flow is also likely to be either too early or disguised by wellbore storage.

Identification of these separate flow regimes is important if the magnitudes of all of the relevant reservoir parameters are to be well understood. Since some of the flow periods usually occur at very early time, minimization of the wellbore storage effect is important to an effective well test. Downhole flow rate measurement and/or downhole shut-off may be used to minimize the influence of wellbore storage. However it must be acknowledged that with the long production interval and common completion practices, there is still likely to be a large wellbore volume on the reservoir side of a downhole flow meter or shut-off valve. Wellbore storage effect may be smaller due to the generally high productivity kh/μ in horizontal wells, but is impossible to remove altogether.

7.1.2 Estimating Parameters

During the early time radial flow period, we can expect to see a semilog straight line behavior with slope m_1 given by:

$$m_1 = 162.6 \frac{qB\mu}{k_H L_w \sqrt{\alpha}}$$

(7.1)

Where α is the vertical to horizontal permeability ratio (k_V/k_H) and L_w is the total effective length of the well (notice that this differs from the terminology of Kuchuk (1995) in which L_w is used to represent the half length of the well).

The skin factor can be estimated from:

$$s = 1.151 \left[\frac{p_i - p_{1hr}}{m_1} + 2\log\left(\sqrt[4]{\alpha} + \sqrt[4]{\frac{1}{\alpha}} \right) - \log\left(\frac{k_H \sqrt{\alpha}}{\phi\mu c_t r_w^2} \right) + 3.228 \right]$$

(7.2)

The slope of the semilog straight line m_1 provides an estimate of the geometric mean permeability $\sqrt{k_H k_V} = k_H \sqrt{\alpha}$. Estimating the skin effect requires knowledge of the vertical to horizontal permeability ratio α, although the value of this parameter does not effect the skin factor strongly.

An estimate of the vertical permeability can be found by examining the time at which the top and bottom boundaries are detected (this is the time at which the behavior departs from the early time radial flow). At the time at which the first boundary is detected:

$$k_V = \frac{\phi \mu c_t}{0.000264 \, \pi \, t} \min\left(z_w^2, (h - z_w)^2\right)$$

(7.3)

At the time at which the second boundary is detected:

$$k_V = \frac{\phi \mu c_t}{0.000264 \, \pi \, t} \max\left(z_w^2, (h - z_w)^2\right)$$

(7.4)

During the late time radial flow, another semilog straight line behavior is expected, with slope:

$$m_2 = 162.6 \frac{qB\mu}{k_H h}$$

(7.5)

The skin factor can be found from:

$$s = 1.151\sqrt{\alpha} \, \frac{L_w}{h} \left[\frac{p_i - p_{1hr}}{m_2} - \log\left(\frac{k_H}{\phi \mu c_t (L_w / 2)^2} \right) + 2.527 \right] - s_z$$

(7.6)

where the geometric effect s_z (for $h_D = \sqrt{\alpha} \, h / L_w < 1.25$) is given by:

$$s_z = -2.303 \log\left[\frac{\pi r_w}{h} \left(1 + \sqrt{\alpha}\right) \sin\left(\frac{\pi z_w}{h} \right) \right] - \sqrt{\frac{1}{\alpha}} \frac{2h}{L_w} \left(\frac{1}{3} - \frac{z_w}{h} + \frac{z_w^2}{h^2} \right)$$

(7.7)

In cases in which a linear flow regime is present, a straight line is expected on a plot of pressure vs. square root of time. The slope of this square root time plot will be given by:

$$m_{sqrt} = \frac{8.128 \, qB}{L_w h} \sqrt{\frac{\mu}{k_H \phi c_t}}$$

(7.8)

The skin factor can be estimated from the intercept of the square root straight line at zero time p_{0hr}:

$$s = \frac{L_w k_H \sqrt{\alpha}}{141.2 \, qB\mu} (p_i - p_{0hr}) + 2.303 \log\left[\frac{\pi r_w}{h} \left(1 + \sqrt{\alpha}\right) \sin\left(\frac{\pi z_w}{h} \right) \right]$$

(7.9)

In cases in which a hemiradial flow period is present, another semilog straight line is expected with slope twice that of the early time radial flow line:

$$m_{hemi} = 2 \, m_1 = 162.6 \frac{2 \, qB\mu}{k_H L_w \sqrt{\alpha}}$$

(7.10)

The skin factor can be estimated from:

$$s = 2.302\left[\frac{p_i - p_{1hr}}{m_{hemi}} + \log\left[\left(1 + \sqrt{\frac{1}{\alpha}}\right)\frac{z_w}{r_w}\right] - \log\left(\frac{k_H\sqrt{\alpha}}{\phi\mu c_t r_w^2}\right) + 3.228\right]$$

$$(7.11)$$

One important difficulty in using most of these estimation procedures is uncertainty in knowledge of the well length L_w. Although the drilled length of the horizontal well may known from the drilling record, it is common that a well may produce along only a portion of its length. The producing length may not be continuous, but may be made up of separated flowing segments along the well. The parameter governing the pressure transient response of the well is the *flowing* length, and not the **drilled** length. Therefore it is usually desirable to conduct a production log in the well to obtain a profile of the flow as a function of length.

7.1.3 Nonlinear Regression

Although nonlinear regression is useful in horizontal well test analysis, complications arise due to correlations between parameters. In particular, skin factor is strongly correlated with both the horizontal permeability k_H and the vertical to horizontal permeability ratio α. These correlations increase the uncertainty in the parameter estimates.

As a demonstration of this point, consider the well test data shown in Fig. 7.6. The data were generated with a value of $\alpha = 0.1$ and a skin $s = 2$, but can be matched perfectly with a value of $\alpha = 1.0$ and a skin $s = 1.75$. Although not shown in Fig. 7.6, the data can also be matched perfectly with a value of $\alpha = 0.01$ and a skin $s = 2.06$.

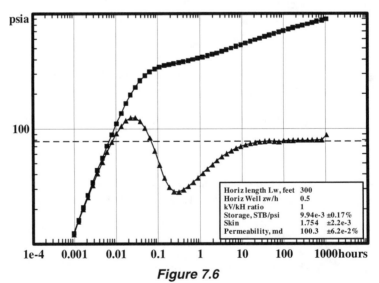

Horiz length Lw, feet	300
Horiz Well zw/h	0.5
kV/kH ratio	1
Storage, STB/psi	9.94e-3 ±0.17%
Skin	1.754 ±2.2e-3
Permeability, md	100.3 ±6.2e-2%

Figure 7.6

One of the reasons that the data in Fig. 7.6 give rise to such uncertainty in the estimate of vertical to horizontal permeability ratio α is that the early time flow regimes are hidden by wellbore storage. This is not uncommon in real tests. Since the time at which the early time radial flow period finishes is almost the only parameter uniquely dependent on vertical permeability k_V, it is practically impossible to determine the vertical to horizontal permeability ratio α unless the early time period is present. For this reason downhole flow rate measurement and downhole shut-off may be required, as described by Domzalski and Yver (1992). Kuchuk et al (1990) recommended that a horizontal well should be tested with two sequential tests, a short drawdown followed by a long buildup.

As with the graphical estimation methods, nonlinear regression is dependent on knowledge of the effective wellbore length. The parameter governing the pressure transient response of the well is the *flowing* length, and not the ***drilled*** length. Therefore it is desirable to conduct a production log in the well to obtain a profile of the flow.

7.2 Multilayered Well Analysis

Most oil and gas wells produce from sedimentary formations, which by their nature are stratigraphic. Therefore it is of interest to determine the effects of the separate layers on the pressure transient observed during a well test. In some cases the layers have no discernible influence, in other cases there is an effect that changes the appearance of the well test data.

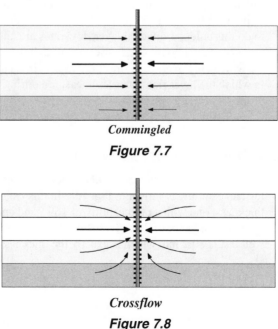

Commingled

Figure 7.7

Crossflow

Figure 7.8

There are two principle types of reservoir layering, depending on whether the layers are in pressure communication with each other within the formation. If the layers are hydraulically separated within the formation (by shales for example) then they are connected only by the open wellbore -- this reservoir model is known as **commingled** layers as in Fig. 7.7. If the layers are in hydraulic communication within the reservoir (as well as through the wellbore) the layers are said to experience **crossflow** as in Fig. 7.8. The pressure transient responses of these two flow models are slightly different.

If we consider the boundary condition on the flowing well for a particular layer j, in the absence of a skin effect the individual layer flow rate q_j can be written:

$$q_j B = -\frac{k_j h_j}{141.2\,\mu_j}\left[r\frac{\partial p}{\partial r}\right]_{r=r_w}$$

(7.12)

From this equation it can be seen firstly that the flow rate from each layer will be proportional to the layer transmissivity kh/μ, and secondly that since the pressure in the wellbore is the same for each layer then the pressure gradient in the formation will also be the same for each layer. Assuming the pressures were the same in all layers before flow to the wellbore started, then in the absence of skin effect there will be no pressure difference between layers at any time -- whether the well is commingled or has formation crossflow. The principle difference between commingled and crossflow models therefore is seen to originate only in the presence of a skin effect.

Much of the interest in multilayered well test analysis is stimulated by the desire to estimate individual layer permeabilities and skin factors. However, it is impossible to do this unless individual layer flow rates are known. Layered well test analysis therefore falls into two categories: (a) tests in which only total flow rate is known, for which only average permeability and skin factor can be estimated, and (b) tests in which the layer flow rates are measured during the test, for which layer properties can be estimated.

7.2.1 Tests without Flow Measurements

A test with multiple layers frequently looks no different from a normal single layer test; an example is shown in Fig. 7.9. In this example there are two layers of equal thickness, one with a permeability of 169.9 md and the other with a permeability of 8.495 md (a factor of 20 difference). The permeability estimate from a match to the well test is 89.2 md, which is exactly the thickness-weighted average of the two.

$$\bar{k} = \dfrac{\displaystyle\sum_{j=1}^{n} k_j h_j}{\displaystyle\sum_{j=1}^{n} h_j}$$

(7.13)

In this case, $0.5(169.9 + 8.495) = 89.19$.

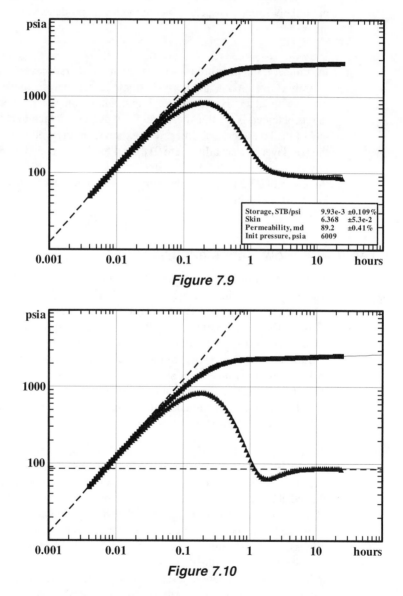

Figure 7.9

Figure 7.10

In other cases in which one layer is much thicker than the other, or where one layer has a larger skin than the other, the response can look like a dual porosity response.

Fig. 7.10 shows an example where one layer has a permeability of 9182 md and a thickness of 0.23 feet and the other has a permeability of 9.182 md (a factor of 1000 difference) and a thickness of 23 feet (a factor of 100 difference). The thick low permeability zone also has a porosity twice that of the thin high permeability zone, and the high permeability zone has a smaller skin factor. Even with these significant contrasts, the deviation of the well test from uniform single layer response is extremely modest. This kind of behavior can only be seen in formations with crossflow, a commingled reservoir will not show such an effect except perhaps in unlikely situations such as where the layers produce from different drainage areas.

In summary, the pressure transient response of a multilayered well usually looks entirely conventional, and normal (single layer) analysis methods can be used to interpret the data. However, it is important to note that the permeability estimate that is obtained is the thickness-weighted average, which may be very different from any of the layer permeabilities. For example, in the well test shown in Fig. 7.10, the two layer permeabilities are 9182 md for 0.23 feet and 9.182 md for 23 feet, however the permeability estimated from the well test is 100 md using a total thickness of 23.23 feet. To determine individual layer properties, it is necessary to know the individual layer flow rates.

7.2.2 Tests with Flow Rate Measurement

Testing of multilayered reservoirs has been a focus of innovation in the field of well test analysis over the past decade or so. Earlier work considered special cases in which traditional approaches had been found to work, however starting with the work by Dogru and Seinfeld (1979) the inherent ill-posedness came to be understood. Prijambodo, Raghavan and Reynolds (1985) and Ehlig-Economides and Joseph (1987) investigated the effects of crossflow between layers and the influence of individual layer skin. Kuchuk, Karakas and Ayestaran (1986) described the application of nonlinear regression to the interpretation of multilayered well tests, but added the important emphasis on the need to determine individual layer flow rates.

Since it is impractical to measure flow rates continuously at all depths, Kuchuk, Karakas and Ayestaran (1986) proposed a testing method in which the well is flowed at a series of different (wellhead) rates with the flow rate measured at the top of a different layer unit in each period. The separate layer responses are then combined using convolution, and the total response matched using nonlinear regression. This approach was developed further by Ehlig-Economides (1993) who described the desuperposition of the individual flow rate changes to make the individual layer responses resemble normal single layer behavior. This approach is similar in concept to rate-normalization as described in "3.7 Desuperposition" on page 97. The procedure can be used either for graphical analysis or to provide first estimates for nonlinear regression.

7.3 References

Dogru, A.H., and Seinfeld, J.H.: "Design of Well Tests to Determine the Properties of Stratified Reservoirs", paper SPE 7694 presented at the 1979 SPE Symposium on Reservoir Simulation, Denver, CO, Feb. 1-2 (1979).

Domzalski, S., and Yver, J.: "Horizontal Well Testing in the Gulf of Guinea", *Oil Field Review*, (April 1992), 42-48.

Ehlig-Economides, C.A., and Joseph, J.: "A New Test for Determination of Individual Layer Properties in a Multilayered Reservoir", *SPE Formation Evaluation*, (Sept. 1987), 261-271.

Ehlig-Economides, C.A.: "Model Diagnosis for Layered Reservoirs", *SPE Formation Evaluation*, (Sept. 1993), 215-224.

Kuchuk, F.J.: "Well Testing and Interpretation for Horizontal Wells", *J. Pet. Tech.*, (Jan. 1995), 36-41.

Kuchuk, F.J., Goode, P.A., Brice, B.W., Sherrard, D.W., and Thambynayagam, M.: "Pressure Transient Analysis for Horizontal Wells", *J. Pet. Tech.*, (Aug. 1990), 974-984; *Trans.*, AIME, **289**.

Kuchuk, F.J., Karakas, M., and Ayestaran, L.: "Well Testing and Analysis Techniques for Layered Reservoirs", *SPE Formation Evaluation*, (Aug. 1986), 342-354.

Prijambodo, R., Raghavan, R., and Reynolds, A.C.: "Well Test Analysis for Wells Producing Layered Reservoirs with Crossflow", *Soc. Pet. Eng. J.*, (June 1985), 380-396.

8. WORKED EXAMPLES

8.1 Drawdown Test Example

The drawdown test data in Example 1 have been analyzed simply in "2.6.1 Semilog Analysis -- Illustrative Example" on page 25 and in "2.7.1 Type Curve Analysis -- Illustrative Example" on page 28. The traditional analyses shown in those earlier sections served to illustrate the concepts under discussion but did not follow the path of a modern analysis that would normally include consideration of the pressure derivative as well as nonlinear regression.

This example illustrates the methodology of a well test interpretation, considering simple tests in homogeneous reservoirs. This kind of response is the most straightforward and "standard" behavior -- however, it is not of purely academic interest, since many real well tests do behave in this manner. More complex cases will be considered in later examples.

In this example well test, a well was originally stable and shut in, within a reservoir at equilibrium. The well was then opened to flow at a constant rate, and the pressure drawdown monitored as a function of time. This is the kind of situation that would be seen in the initial production of a discovery well in an otherwise unexploited field (although the data in this example have been generated artificially to demonstrate particular issues).

In a traditional graphical interpretation of a well test (in a simple, homogeneous reservoir) the following steps are commonly used:

(a) Identify the various flow periods by looking at the data on derivative, and semilog plots.

(b) Estimate individual reservoir parameters from the log-log and semilog plots.

(c) Perform a type curve match to the log-log derivative type curve, and again estimate the reservoir parameters, checking for consistency with the prior estimates.

These same steps are likely to be taken as the preliminary stage of a computer-aided interpretation, as a means of evaluating the data, choosing a reservoir model and making a first estimate of the reservoir parameters.

8.1.1 Identifying Flow Periods

To begin with the first step, the data are displayed in a log-log presentation of the pressure Δp and derivative $t\, dp/dt$ (Fig. 8.1).

In the first log cycle, there is a clear indication of wellbore storage effect (unit slope). In the third log cycle, the derivative flattens out, indicative of infinite acting radial flow. The second log cycle, which lies between the two, is a transition period -- the "hump" in the pressure derivative is a feature that almost always comes at the end of a storage response. There are no other reservoir responses indicated -- no boundary effects after the infinite acting period, and no dual porosity effects.

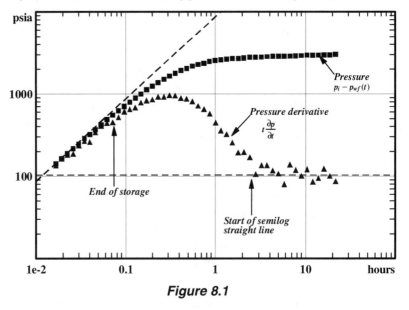

Figure 8.1

Before accepting the suggested interpretation model (storage followed by infinite acting radial flow), it is essential to make a consistency check. If the flattening region or the derivative is truly indicative of the current semilog straight line, then it should begin one and a half log cycles after the end of the unit slope period of the storage response. A check of Fig. 8.1 confirms that this is so. Hence we accept the preliminary evaluation of the reservoir model, and note for future use that the unit slope storage response ends at about 0.1 hours and the correct semilog straight line (infinite acting radial flow) begins at about 2 hours. This attention to recognizing the start of the infinite acting period is important, since finding the **correct** semilog straight line is critical to achieving valid estimates of permeability, skin and reservoir pressure.

8.1.2 Estimating Parameters

Moving on to the second step, the reservoir parameters are estimated. Known parameters in this example are as follows:

B_t (RB/STB)	1.21	r_w (feet)	0.401
μ_o (cp)	0.92	h (feet)	23
c_t (/psi)	8.72×10^{-6}	p_i (psia)	6009
ϕ	0.21	q (STB/d)	2500

It is convenient to estimate the wellbore storage first, since this will serve again as a more precise indication of the time that wellbore storage ends. Fig. 8.1 shows a unit slope line drawn through the early data.

From the equation of the unit slope straight line

$$(p_i - p) = \frac{0.234}{5.615} \frac{qB}{C} t \qquad (8.1)$$

the storage coefficient C can be estimated from **any** point on the line.

The estimated value of storage coefficient, C, is 0.0154 STB/psi. In Fig. 8.1 it appears that the data begin to move away from the unit slope line rather sooner than 0.1 hours. This emphasizes the fact that data in this storage period are measured only over a five minute time frame, and therefore, as discussed earlier, will be sensitive to time datum errors in the measured pressure data. This will be discussed later. Notice also that the appearance of the data is governed by the numerical value of initial reservoir pressure p_i. For this drawdown test, the value of p_i is the pressure measured downhole just before the well was opened to flow.

Moving on to the semilog graph of pressure $p_{wf}(t)$ against time (Fig. 8.2), a semilog straight line portion of data (starting at about 2 hours as expected) can be seen.

The slope of this line, m, is 255.2 psi/log cycle, which allows us to estimate permeability, k, using Eq. 2.29:

$$m = 162.6 \frac{qB\mu}{kh}$$

$$255.2 = 162.6 \frac{(2500)(1.21)(0.92)}{k(23)}$$

Hence the permeability $k = 77.1$ md.

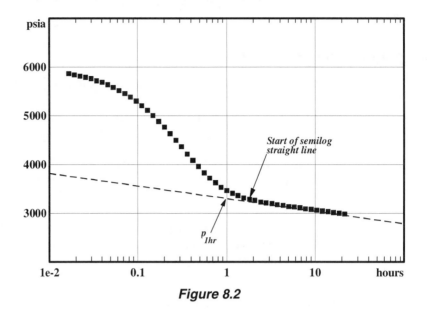

psia

Figure 8.2

The skin factor, s, can be estimated from the intercept of the line, using Eq. 2.30:

$$s = 1.151\left[\frac{p_i - p_{1hr}}{m} - \log\frac{k}{\phi\mu c_t r_w^2} + 3.2274\right]$$

$$s = 1.151\left[\frac{6009 - 3330}{255.2} - \log\frac{77.1}{(0.21)(0.92)(8.72\times10^{-6})(0.41)^2} + 3.2274\right]$$

Hence the skin factor $s = 6.09$.

It is important to note that the point p_{1hr} is a point **on the semilog straight line** and may not be a point in the data itself (in Fig. 8.2 the actual data point at time one hour [3465 psi] lies above the point on the semilog straight line used to determine the skin factor [3330 psi]).

At this point, we have used all of the identifiable periods (i.e., storage period, infinite acting period), and thus there are no further parameters than can be estimated. It is noted that the data lying in the transition period (the 1½ log cycles from 0.1 hours to 2 hours) have not been used in any quantitative way. To use this section of the data we could do a type curve analysis that matches the entire data set to the reservoir model, including the transition data as well.

8.1.3 Type Curve Match

Even though the reservoir parameters have already been estimated, there are several advantages in performing a type curve match. Whereas the semilog method and unit slope log-log line used only portions of the data, a type curve match uses the entire data set. This helps ensure consistency over the whole range of time, and also provides a mechanism to make use of the transition data which lies between the individual response periods.

In a log-log type curve, it is known that the p_D versus t_D curves (the reservoir model) will have exactly the same shape as the p_i - p_{wf} versus t data (the measurements during the well test). Thus the data are plotted on transparent three inch by three inch log-log graph paper (as in Fig. 8.1), to the same scale as the three inch log cycle type curve. The data are then laid over the type curves, and moved horizontally and vertically (keeping the axes parallel) until a match is achieved, as in Fig. 8.3.

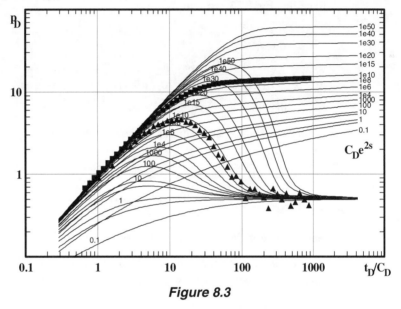

Figure 8.3

The inclusion of the pressure derivative in the type curves and in the data makes the matching process much easier than was shown in "2.7.1 Type Curve Analysis -- Illustrative Example" on page 28, where the derivative was not included.

At the point of matching, correspondence between p_D and Δp and between t_D and t has been achieved and the reservoir parameters can be estimated as described in "2.7 Log-Log Type Curves" on page 27. From Fig. 8.3 it is also possible to estimate skin factor s, by noting which of the various curves gives the best match, and reading the value of $C_D e^{2s}$ appropriate to that curve.

Looking at Fig. 8.3, we can choose **any** point to represent the correspondence between dimensional and dimensionless properties. For example, choosing the last data point at t = 21.6 hrs, p_i - p_{wf} = 6009 - 2988.93 = 3020.07 psi, we can read the corresponding values on the dimensionless axes to be t_D/C_D = 800 and p_D = 13.67.

Using the pressure match:

Recalling Eq. 8.32, we find:

$$\log \Delta p = \log p_D - \log \frac{kh}{141.2qB\mu}$$

$$\log(6009 - 2988.93) = \log 13.67 - \log \frac{k(23)}{141.2(2500)(1.21)(0.92)}$$

Hence, the permeability $k = 78.15$ md.

Using the time match:

Recalling Eq. 8.33, we find:

$$t_D = \frac{0.000264kt}{\phi\mu c_t r_w^2}$$

$$800C_D = \frac{0.00264(78.15)(21.6)}{(0.21)(0.92)(8.72 \times 10^{-6})(0.41)^2}$$

Hence, the dimensionless storage coefficient $C_D = 1960$.

From the definition of the dimensionless wellbore storage coefficient (Eq. 2.17):

$$C_D = \frac{5.615C}{2\pi\phi c_t h r_w^2}$$

$$C = 1960 \frac{2\pi(0.21)(8.72 \times 10^{-6})(23)(0.41)^2}{5.615}$$

Hence, the storage coefficient C=0.0155 STB/psi.

We can also see that the best curve matched is at a value of $C_D e^{2s} = 5 \times 10^8$, thus we can estimate skin factor s using the previously estimated value of C_D:

$$e^{2s} = (5 \times 10^8)/1960$$

Hence, the skin factor s=6.22.

8.1.4 Comparison of Graphical Estimates

It is interesting to compare the estimates of the parameters from the semilog methods and from the type curve matching procedures, as shown in Table 8.1.

Parameter	Semilog Estimate	Type Curve Estimate
k (md)	77.1	78.15
s	6.09	6.22
C (STB/psi)	0.0154	0.0155

Table 8.1

In this case the estimates are extremely close, and therefore we gain confidence in our interpretation since we have demonstrated consistency between the semilog and log-log methods. Actually in this example the consistency is considerable better than would ever be possible with pencil and paper graphical methods, since the estimation here actually used computer-aided graph plotting and slope finding algorithms. It is a rather simple task for a computer to place the type curve match in exactly the location set by the slope fitting methods -- the human interpreter need only make a decision as to whether the match is good or not.

Log-log type curve matching is not as precise as semilog methods, since the log-log axes tend to hide inaccuracies at late time (large pressure drop). For example, if we calculate the model response using the recent estimates of the parameters, then we can compare this "customized" type curve with the data, as in Fig. 8.4.

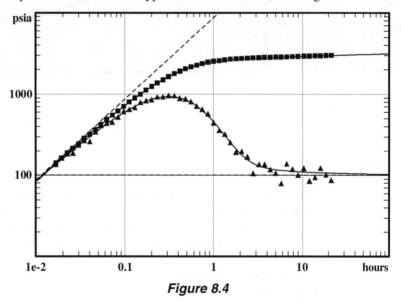

Figure 8.4

This step is sometimes referred to as **simulation** (not to be confused with numerical reservoir simulation) in that we have *simulated* the way the well test would have looked if our estimates were the correct answer.

Plotting the simulated test in log-log coordinates, as in Fig. 8.4, makes the match appear quite good. However, it must be remembered that a 1mm deviation of a data point from the (pressure) curve at late time means an actual error of about 200 psi, whereas the same deviation at early time means an actual error of only 5 psi. Thus a visually inspected type curve match in log-log coordinates tends to be imprecise at the place we need greatest definition, that is during the actual reservoir response. Plotting derivatives is a considerable help, since the derivative plot tends to lie in the region of the log scale with greatest precision. Even so, examining the same simulation in semilog coordinates (Fig. 8.5) shows that it is not such a good match after all, especially in the transition leading into the semilog straight line region.

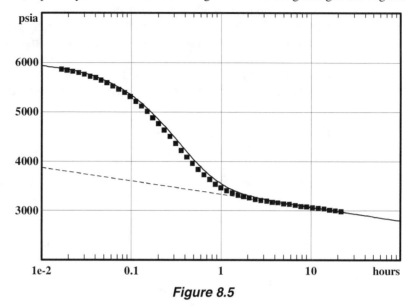

Figure 8.5

8.1.5 Time Shift Errors

As discussed previously in "3.5.2 Datum Shifting" on page 87, errors in registering the time at which the flow started can affect the early time data considerably. In this particular example, the storage period lasts only until about six minutes into the test, so a time error of only one minute can change the appearance of the log-log graph and hence the estimate of the wellbore storage coefficient. Fig. 8.6 shows the appearance of the data if time zero is shifted one minute earlier and Fig. 8.7 shows the appearance if it is shifted one minute later.

The unit slope straight line drawn in Figs. 8.6 and 8.7 is in the same location as that estimated for the original (correct) data, showing clearly how difficult it would be to correctly identify this line if the time datum was only one minute in error.

Luckily, the potential difficulty of time shift is overcome rather easily. Instead of insisting on that critical first data point ($t = 0$, $p = p_i$ in this example) we can ignore the measured value and *calculate* pressure p at time $t = 0$ by extrapolating back the subsequent data points. Removing this one single data point makes the data appear much more conventional on the log-lot plot, Fig. 8.8, even though the slightly incorrect data is still being used.

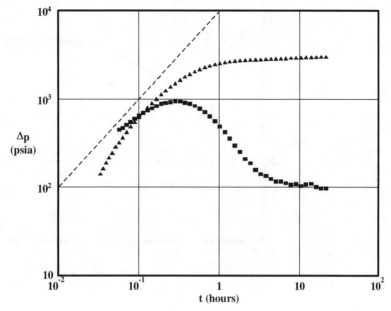

Figure 8.6

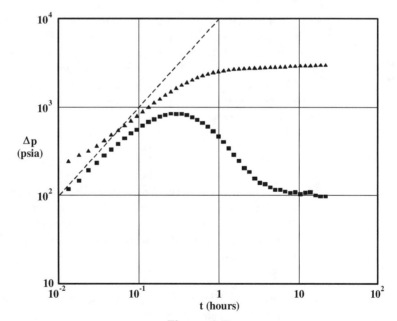

Figure 8.7

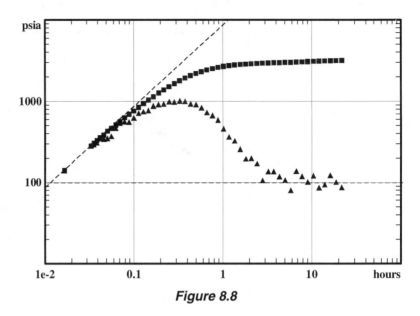

Figure 8.8

8.1.6 Pressure Shift Errors

As with time shift, an error in the initial pressure value used to compute Δp can result in aberrations in the log-log diagnostic plot. For example, Fig. 8.9 shows the effect on the plot if the initial pressure value is 100 psi in error. Although pressure can be measured with much greater precision than 100 psi, there often remains a difficulty in determining the pressure at the start of the test since it may not be clear exactly when the transient began, and pressures change very rapidly and very much during this early period.

The difference between the effects of time shift errors and pressure shift errors can be seen by comparing Figs. 8.7 and 8.9. Although these two plots look similar at first glance, on closer examination it can be seen that the time shift error (Fig. 8.7) affects both the pressure and the pressure derivative, while the pressure shift error (Fig. 8.9) affects only the pressure (the derivative still follows the expected unit slope behavior, since the derivative is a function of pressure **differences** only and not of actual pressure values).

As with time shift error, to overcome pressure shift errors we can ignore the measured value and **calculate** pressure p at time $t = 0$ by extrapolating back the subsequent data points.

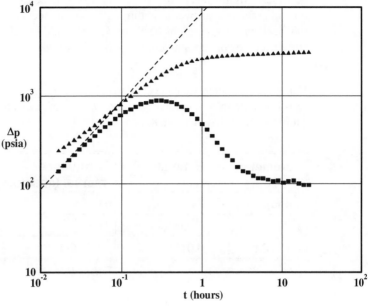

Δp (psia)

t (hours)

Figure 8.9

8.1.7 Confidence Intervals

The estimates obtained by graphical analysis (as listed earlier in Table 8.1) are about as close to the correct answer as a graphical analysis by hand is ever likely to come (since the test data in this example were artificially generated, the correct answer is actually known!). In fact, having used computer-assisted slope finding methods, these may be even better than can be achieved by eye. Nonetheless, using pencil and paper, different engineers will certainly obtain different estimates. Table 8.2 lists twenty different estimates of permeability obtained for this problem by a class of twenty petroleum engineering students. Discarding the highest and lowest values as probable outliers, the remaining estimates have a variation of about $\pm 10\%$.

80.10	81.52	73.43	78.11	82.35
75.07	75.35	70.08	82.76	77.44
74.44	76.93	79.86	75.31	87.86
75.11	76.71	76.14	81.81	74.34

Table 8.2

Thus within this human group, there is an uncertainty of $\pm 10\%$ as to the correct value of permeability. Note also that this is a rather clear example with good data -- with noisier data in a more complex situation the range of estimates will probably

---

be wider. All this means that if a supervisor wanted to be reasonably sure of obtaining the best answer to within 10%, 20 engineers could be ordered to perform independent interpretations. This of course would be ridiculously expensive, and still the degree of certainty could not be quantified.

Using the computer, we can calculate 95% confidence intervals, as described in "3.6.1 Confidence Intervals" on page 89. The intervals for our best graphical estimates are listed in Table 8.3.

Parameter	Estimate	Confidence Interval (absolute)	Confidence Interval (percent)
k (md)	77.1	±11.84	±15.36%
s	6.09	±1.890	±31.04%
C (STB/psi)	0.0154	±0.5666×10⁻³	±3.680%

Table 8.3

Within the criteria defined in Section 3.6.1 (Table 3.2) the confidence intervals are estimates of permeability, k, and skin factor, s, are beyond acceptable limits. Thus we cannot be sure (to within 95% confidence) that this answer is acceptable.

8.1.8 Nonlinear Regression

Using nonlinear regression, the computer can obtain the best fit to the data, using the reservoir model that we have now identified (storage and skin, infinite acting reservoir, no boundary effect). The procedure requires a preliminary estimate of the unknown reservoir parameters, so the values obtained previously from the graphical analysis will be used. A (semilog) plot of the final match, Fig. 8.10, shows that a much closer match has been obtained, and Table 8.4 shows that the confidence intervals are now much narrower.

Parameter	Estimate	Confidence Interval (absolute)	Confidence Interval (percent)
k (md)	86.49	±1.046	±1.21%
s	7.690	±0.1687	±2.19%
C (STB/psi)	0.01453	±0.3718×10⁻⁴	±0.26%

Table 8.4

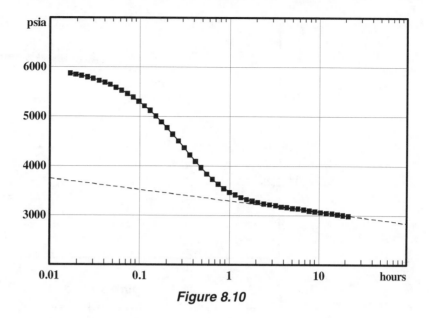

Figure 8.10

Even though nonlinear regression is sometimes known as automated type curve matching, it is important to recognize that the two are quite different. Nonlinear regression does not depend on log-log curve matching, and will minimize the deviation of the reservoir model from the data by an equivalent amount over all time ranges.

It might reasonably be questioned how nonlinear regression is affected by the time shifting error described earlier in "8.1.5 Time Shift Errors" on page 152. Analyzing the data of Fig. 8.7 (which had a one minute time shift) results in a poor match, Fig. 8.11, although the wide confidence intervals, Table 8.5, warn us that the estimated answer is not acceptable.

Parameter	Estimate	Confidence Interval (absolute)	Confidence Interval (percent)
k (md)	72.03	± 12.16	$\pm 16.88\%$
s	5.317	± 1.936	$\pm 36.41\%$
C (STB/psi)	0.01276	$\pm 0.6173 \times 10^{-4}$	$\pm 4.84\%$

Table 8.5

The solution to this problem is much as it was in the graphical analysis, the first data point is removed, and the nonlinear regression routine is forced to match the initial pressure, p_i, in addition to k, s and C. The new match is shown in Fig. 8.12. In this case the answer is almost as before, with acceptable confidence intervals on all parameters, Table 8.6.

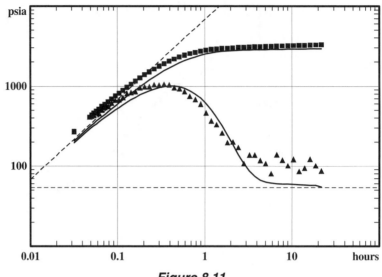

Figure 8.11

Parameter	Estimate	Confidence Interval (absolute)	Confidence Interval (percent)
k (md)	87.54	±1.210	±1.38%
s	7.146	±0.1824	±2.55%
C (STB/psi)	0.01540	±0.7084×10^{-4}	±0.46%
p_i (psia)	5869.	±2.480	±0.04%

Table 8.6

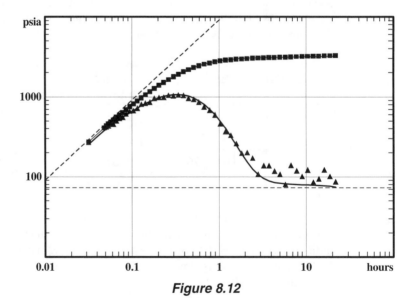

Figure 8.12

Although the permeability, k, estimate is almost the same as before, the storage coefficient, C, estimate might have been further away (although in this example the estimate of C is the same as before). If a different estimate had been obtained, this would be because it is the early time data that is most in error. Fortunately, C is not a parameter that we are actually interested in finding. The estimate of the initial reservoir pressure, p_i, is 5869 psia, whereas the true measured value was 6009 psia. This difference should serve as a warning as to the degree of accuracy to which p_i can be estimated, and the sensitivity of the estimate to the early time data in a single constant flow rate test. For reference, the extrapolated value of p_i that would be obtained graphically would be 5867 psia -- essentially the same value. The skin factor estimate, s, differs from the previous value by about 0.5, due to the change in the apparent pressure drop when the value of p_i changes.

To obtain a more accurate estimate of the reservoir pressure, it is necessary to go back to data preparation, and shift the time values by the small increment required to straighten the unit slope storage line on the log-log graph.

8.1.9 Radius of Investigation

In this example data set there is no evidence of any pressure transient caused by boundary effect. As time passes during the test, the pressure transient response at the well will reflect reservoir conditions at greater radius. Recall from Section 2.8.1 that for a closed circular reservoir, the boundary response begins at a dimensionless time t_{DA} of 0.1. Given that no such response was observable in the example data, we can ascertain that t_{DA} at time $t = 21.6$ hours (the last data point measured) must be smaller than 0.1. From this we can determine that the radius of investigation was less than the actual radius of the reservoir.

Putting this into numbers, the area and radius of investigation can be determined from:

$$t_{DA} = \frac{0.000264\,kt}{\phi\mu c_t A} < 0.1$$

$$\frac{0.000264(86.49)(21.6)}{(0.21)(0.92)(8.76\times10^{-6})A} < 0.1$$

$$A > 2.93\times10^6 \ ft^2$$

$$r_e > 965 \ ft$$

This value is very close to the value of 943 ft estimated earlier for this example in "2.9 Radius of Investigation" on page 35. From this estimate of the radius of investigation, we do not know where the actual reservoir boundaries are, but we have determined that they are at least 965 feet away. We have therefore been able to place an approximate lower bound on the reservoir size. However it should be emphasized that this radius of investigation is a rather broad estimate of how far the test "sees" into the reservoir -- the time at which the well pressure begins to be affected by the boundary may not be the time that the interpreter is able to notice the effect. It is possible that the reservoir could have a boundary closer than 965 feet, but that its influence is not strong enough to be noticeable. In practice, it is expedient to assume that the lower bound on reservoir size is rather smaller than the radius of investigation, perhaps using the value 600 ft in this example.

8.1.10 Truncated Data

As discussed in "3.6 Nonlinear Regression" on page 88, nonlinear regression provides us with the capability to make estimates of reservoir parameters based on data in *transition* periods, thus allowing us to interpret well tests that have been truncated even before reaching the infinite acting semilog straight line. For example, suppose our original drawdown had been cut off after only 1.55 hours by a failure of the pressure tool. In this case we know that the correct semilog straight line does not begin until around two hours, thus it would not be possible to do a traditional semilog analysis to estimate permeability and skin factor. We might be tempted to try a type curve match, however this is not a good idea since it is difficult to obtain a unique match, even with the derivative type curve, if the semilog straight line data are not present. Fig. 8.13 shows a reasonable looking type curve match, even though the parameter estimates are completely incorrect.

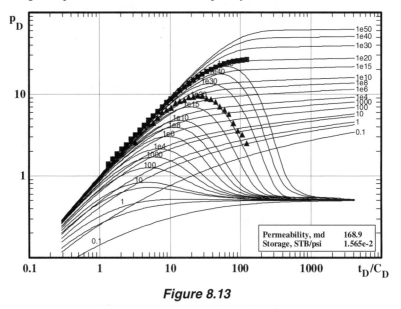

Figure 8.13

Using nonlinear regression, we can obtain a good match to this data (Fig. 8.14) even though it contains only storage dominated and transition data. The confidence intervals are all still within acceptable range (Table 8.7), and we can see from our original interpretation that the answers are correct.

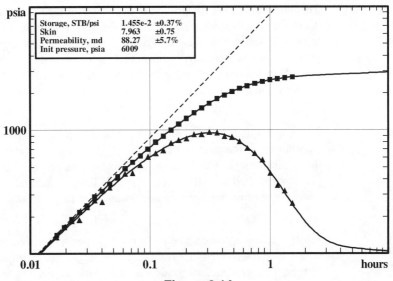

Figure 8.14

Parameter	Estimate	Confidence Interval (absolute)	Confidence Interval (percent)
k (md)	88.27	±5.032	±5.70%
s	7.963	±0.7505	±9.42%
C (STB/psi)	0.01455	±0.5384×10^{-4}	±0.37%

Table 8.7

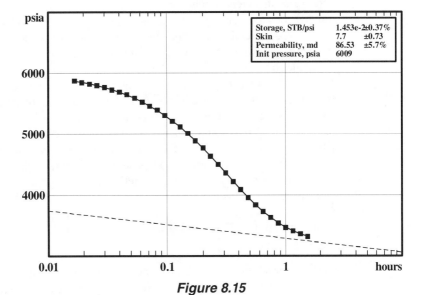

Figure 8.15

The most impressive illustration of the power of the nonlinear regression technique can be seen in Fig. 8.15, which shows the correct semilog straight line drawn in a location based on the estimates of k and s from nonlinear regression. It is clear from this plot that *no part* of the actual data lies on the straight line. Nonlinear regression has been able to correctly ascertain the position of the straight line based on transition data alone.

In this example, nonlinear regression would have allowed us to obtain the correct answer even if the test had been only 1.55 hours long (compared to 21.6 hours in the original test) or would have saved us from having to rerun the test if the pressure tool had actually failed after 1.55 hours. In either of these two situations, the expense of the well testing operation could be reduced substantially.

While discussing truncated data, it needs to be emphasized that not all truncated data can be interpreted in this way. If the data in our example were truncated still further, the confidence intervals would become so much broader that we could no longer rely on the estimate. Also, the objective of the test may have been to identify the reservoir model or to explore for reservoir limits -- in either case the full length of data would still be required. Nonetheless there are many routine well tests in producing fields that could be run for shorter time, if interpreted by nonlinear regression. There are also probably many older misdesigned tests or data sets truncated by tool failure that could be usefully interpreted using this approach.

8.1.11 Simultaneous Rate-Pressure Data

Improvements in measurement technology now make it possible to make simultaneous measurement of pressure and flow rate, although this is still problematic in multiphase wells. Flow rate measurement can either be made directly by a downhole flow meter, or indirectly by monitoring fluid levels in a downhole shut-off or DST tool. The advantage of knowing the sandface flow rate is that the wellbore storage effect is no longer of any consequence, and all of the measured pressure response is that of the reservoir rather than of the well. In general this means that more reservoir information can be obtained earlier in the test. However, some caution is necessary since the radius of influence at early time may be quite small, and the tested part of the reservoir may be only the skin damaged zone close to the well. Thus, although it may be possible in undamaged wells to analyze very short data, the likelihood of good interpretations in wells with skin damage is reduced. Rather than a means of interpreting shorter tests, use of simultaneous rate data should be seen as a way of obtaining more and less ambiguous information from a test of normal length.

To demonstrate the concept, as well as some of the difficulties, consider our same drawdown example test. Fig. 8.16 shows the pressure and flow rate history during the first 15 minutes of the drawdown (the eventual flow rate was 2500 STB/day).

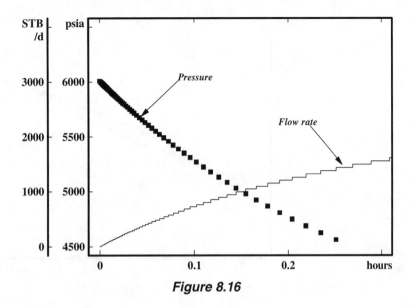

Figure 8.16

Fig. 8.17 shows the same period of pressure data plotted on a log-log graph, showing that the response lies more or less totally within the wellbore storage regime. Variable rate data can be "straightened" by rate-normalization, to show an apparent semilog straight line in Fig. 8.18.

Using nonlinear regression, we can obtain estimates of k and s (C is no longer of significance) that are similar to the previous estimates. Using only 15 minutes of data, the estimates are obtained far from the semilog straight line they represent, Fig. 8.19.

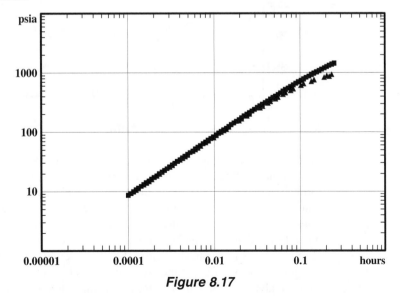

Figure 8.17

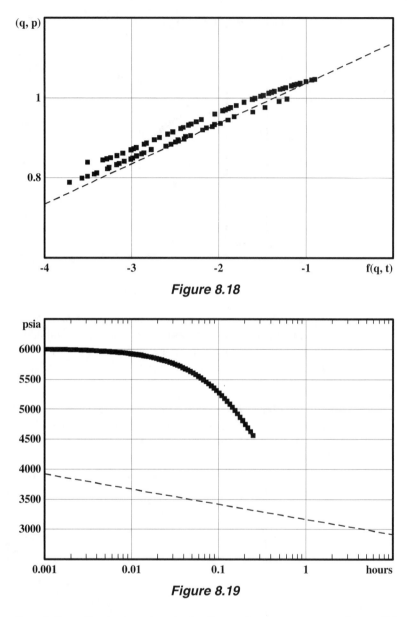

Figure 8.18

Figure 8.19

Even though the estimates are close to the (known) correct answers, the confidence intervals are wider than commonly acceptable. This is because of a very high correlation between k and s over this time interval since the pressure drop across the skin is a significant proportion of the total pressure drop. The radius of investigation during 15 minutes was only about 100 feet. Thus, over this period, the skin effect is dominant, making it more difficult to analyze the reservoir response. Thus in some ways we have avoided the storage effect, only to be trapped by the next non-reservoir effect in the sequence. This limits the extent to which short data can be analyzed, even with flow measurements, although the added advantages of flow measurement over normal test lengths are still major.

8.2 Bounded Reservoir Example

To estimate a boundary location with more precision, it is necessary to run a longer test. Drawdown Example 1(a) includes an additional two days of pressure transient data beyond the original 21.6 hours of flow in Example 1. Examination of the diagnostic plot (Fig. 8.20) reveals that there is an evident boundary effect, based on the late time unit slope behavior of the derivative plot we can say that this is probably a pseudosteady state response. From our diagnosis of the derivative, the boundary effect appears to start at around 25 hours, while radial flow appears to occur between about 2 hours and about 20 hours.

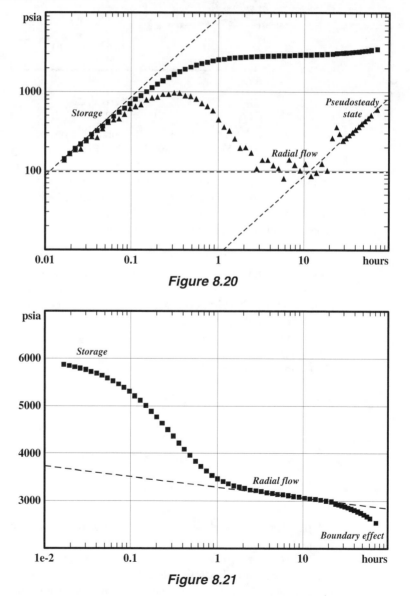

Figure 8.20

Figure 8.21

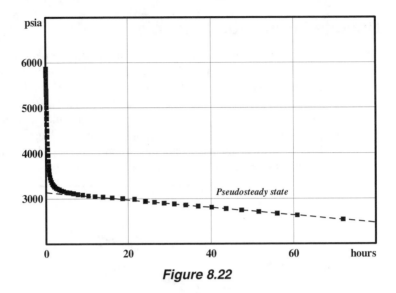

Figure 8.22

Notice that the behavior is much more noticeable on the derivative plot than on the pressure plot. Looking at the semilog plot (Fig. 8.21), we again see the boundary effect more clearly than on the log-log (pressure) plot and also that the boundary response does not seem like that of a fault (doubling in slope). Based upon this observation, and on the earlier derivative view, we look at a Cartesian plot of pressure against time (Fig. 8.22) which does show a straight line behavior characteristic of pseudosteady state, starting at the expected time of about 25 hours.

It is important to note that it is easy to find "straight lines" on a Cartesian plot -- even when none are present. (For example, it is not difficult to imagine a straight line on a Cartesian plot, even with the original 21.6 hours of data). Therefore it is important that the pseudosteady state be clearly identified on the derivative and semilog plots, before it is drawn on the Cartesian plot. After observing the slope of the Cartesian straight line, $m_{Cartesian}$, to be 8.54 psi/hr the area of the reservoir can be estimated using Eq. 2.39:

$$m_{Cartesian} = \frac{0.2342qB}{(\phi c_t h)A}$$

(8.2)

$$A = \frac{0.2342\,qB}{\phi \mu c_t h m_{Cartesian}} = \frac{0.2342(2500)(1.21)}{(0.21)(8.72 \times 10^{-6})(23)(8.54)}$$

$$A = 1.97 \times 10^6 \ ft^2; \quad r_e = 792\,ft$$

Notice that this is slightly less than the radius of investigation estimated previously (965 ft), emphasizing that it would have been wrong to treat 965ft as a lower bound on reservoir size, since we have now observed from the later data that the boundary response was already starting, except that it was not prominent enough to be noticeable. Checking the estimate in a simulation step confirms that the r_e estimate of 792 ft is consistent, and nonlinear regression refines the estimate (Fig. 8.23) and demonstrates that the confidence limits are within acceptable bounds (Table 8.8).

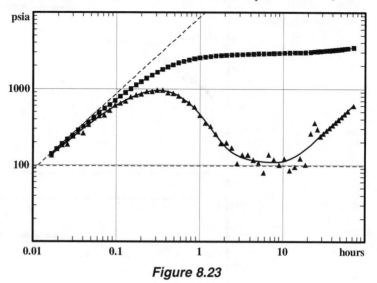

Figure 8.23

Parameter	Estimate	Confidence Interval (absolute)	Confidence Interval (percent)
k (md)	88.62	±1.759	±1.98%
s	8.020	±0.2832	±3.53%
C (STB/psi)	0.01456	±0.5702×10^{-4}	±0.39%
p_i (psia)	794.2	±5.541	±0.70%

Table 8.8

The estimates of permeability and skin have changed somewhat, as we can see from the derivative plot (Fig 8.23) that the last part of the original semilog straight line has now been attributed to boundary effect. The confidence intervals have also grown slightly wider -- this is due to the fact that four parameters were matched, compared to the original three. It is generally not a good idea to place too much importance in relative sizes of confidence intervals, instead we are interested principally in whether they are within the acceptable range or not.

Before concluding our discussion of boundary effects we can compare our final interpretation with the responses that would have been observed under other reservoir boundary conditions. Fig. 8.24 shows the simulated responses that would have been seen for impermeable faults and constant pressure boundaries. Again notice how much easier the response is to recognize on the derivative plot, compared to the log-log pressure plot.

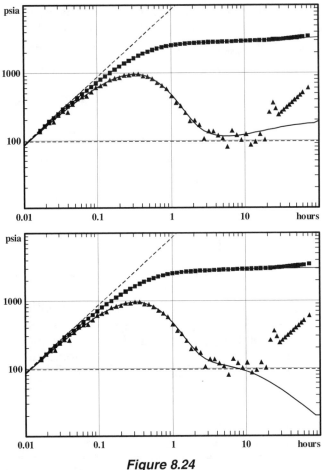

Figure 8.24

8.2.1 Ambiguity

In most well test analyses there is room for more than one interpretation, since the appearance of the data may be **ambiguous**. In many cases ambiguity is not a problem, since usually only one interpretation will be consistent with other known information (e.g., geology). In addition, investigation of the confidence intervals provides a useful tool for the recognition of ambiguity.

As an example, even the simple drawdown case considered in the preceding sections can permit alternative interpretations (with some imagination). Looking at the original 21.6 hour data set in a semilog plot (shown previously in Fig. 8.2), we might imagine that the semilog straight line is already due to the boundary effect of an impermeable fault, and that its slope is in fact double that of the infinite acting semilog straight line (which would then be hidden in the storage dominated period). In this case the semilog interpretation would look as in Fig. 8.25.

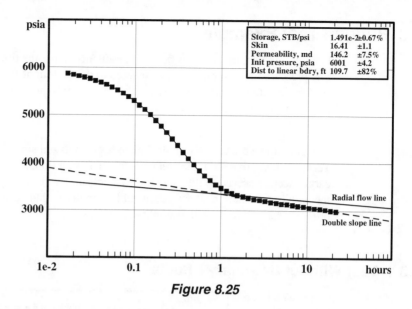

Storage, STB/psi	1.491e-2	±0.67%
Skin	16.41	±1.1
Permeability, md	146.2	±7.5%
Init pressure, psia	6001	±4.2
Dist to linear bdry, ft	109.7	±82%

Figure 8.25

Starting with these estimates, we can obtain a very close looking match, also shown in Fig. 8.25, by nonlinear regression (if we start with the original estimates, nonlinear regression will force the supposed impermeable fault beyond the radius of investigation and converge to the same infinite acting solution as before).

Fortunately, the confidence intervals (shown in Fig. 8.25) warn us that the solution is not statistically significant, since the confidence interval on the fault distance is so large (82%) even though we have already ascertained that there is a long semilog period (which is all boundary effect under the interpretation being considered).

The reason for the wide confidence interval is that, without any infinite acting period, the fault boundary could be at almost any location within the radius of investigation at the end of the storage period (even right at the well itself). On the basis of these unacceptable intervals, we would reject the interpretation. This is fortunate, since we would otherwise be in error (by a factor of two) in our estimates of permeability and skin factor. Consideration of the confidence intervals, and recognition of the correct starting point of the infinite acting semilog straight line, has rescued us from an invalid interpretation.

Before concluding, it is important to reiterate that true ambiguity in well test interpretation is less common than sometimes believed. Although some mathematical models of reservoir behavior give rise to similar looking responses, the process of a well test interpretation involves more than just mathematics. Frequently, information from geology and well logs can be used to resolve an apparent ambiguity.

8.3 Buildup Test Example

In a buildup test, the well has been on production for some time prior to the test. The pressure measurement tool is lowered to the bottom of the well, and the well is shut in (in actuality the well may be shut in to install the pressure tool, then restarted, stabilized, and shut in again).

The buildup test data in Example 2 have been analyzed simply in "2.15.1 Buildup Tests -- Illustrative Example" on page 58. The traditional analysis shown in that earlier section served to illustrate the concepts under discussion but did not follow the path of a modern analysis that would normally include consideration of the pressure derivative as well as nonlinear regression.

8.3.1 Test Without Boundary Effects

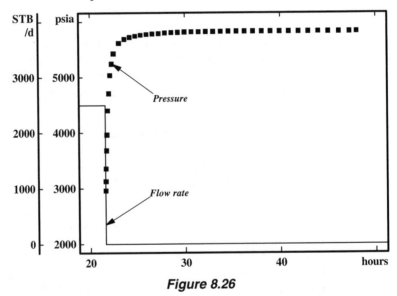

Figure 8.26

Fig. 8.26 shows the pressure and flow rate for Example 2. If we do a standard diagnosis using a (drawdown) derivative plot, Figure 8.27, we see an apparent downward trend in the derivative suggestive of a boundary effect. Nonetheless this downward trend is due to the buildup effect since the producing time (21.6 hours) is short relative to the shut-in time. Making a plot using Agarwal effective time in Fig. 8.28 removes this apparent downward trend and shows a more conventional behavior. Based on the apparent start of the infinite acting radial flow behavior (at a shut-in time of around 2 hours, we can look for the start of the Horner straight line to be at a Horner time of about 11.8. Fig. 8.29 shows a Horner straight line starting at the expected place. This is the line that was used in the analysis in "2.15.1 Buildup Tests -- Illustrative Example" on page 58.

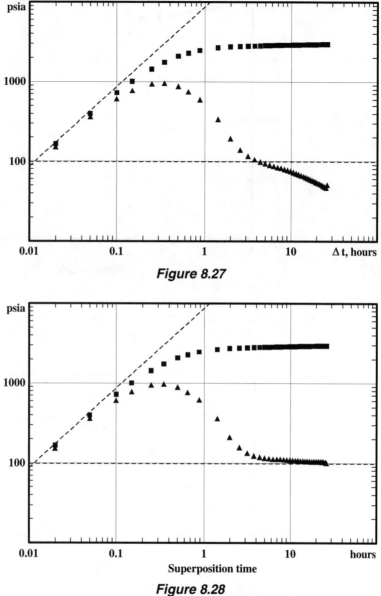

Figure 8.27

Figure 8.28

It may not always be correct to perform derivative diagnosis in this way, since the effects of the transient prior to the shut-in can change the appearance of the derivative plot considerably. Traditionally, it has been assumed that drawdown type curves can be used for buildup if the shut in time is no longer than 20 to 30 percent of the producing, but this may not be valid if boundaries are present.

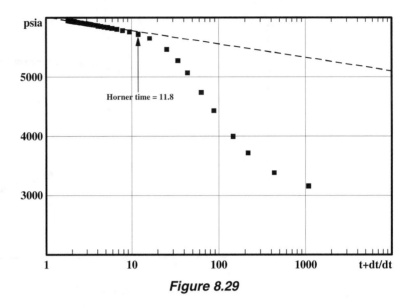

Figure 8.29

However, the example shown here shows no major distortions in shape (provided effective time or superposition time is used), because there is no influence of any boundaries during the shut-in *nor during the preceding production*.

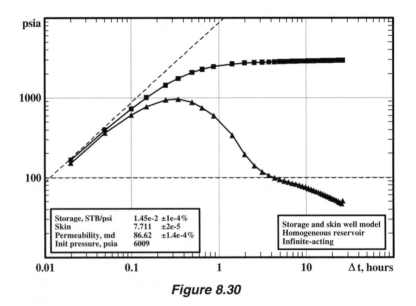

Storage, STB/psi	1.45e-2	±1e-4%
Skin	7.711	±2e-5
Permeability, md	86.62	±1.4e-4%
Init pressure, psia	6009	

Storage and skin well model
Homogeneous reservoir
Infinite-acting

Figure 8.30

Nonlinear regression may be used to match the buildup data in the same way as drawdown data. Unlike Horner analysis or the superposition time approach used to visualize the derivative plot (both of which are based on an assumption that both the drawdown and buildup responses are in infinite acting radial flow) there is no approximation involved in nonlinear regression. During the matching process, the nonlinear regression approach computes the pressure response irrespective of the reservoir model, and includes whatever flow rate history preceded the shut-in. In this way, nonlinear regression provides a much more conclusive interpretation of the data.

Fig. 8.30 shows a nonlinear regression match to this data, plotted on a normal log-log plot to illustrate the point that the mathematically computed response also shows the downward trend even for an infinite acting model.

8.3.2 Effects of Boundaries

Understanding the influence of boundaries is a key issue in buildup tests. The measured response is characteristic not only of the time of the buildup itself, but also of the preceding drawdown. Thus the duration of the shut-in period may be much less than is required to reach a boundary response, but the drawdown transient, which has continued for much longer (and effectively continues during the buildup too), may be influenced by the drainage boundaries. This boundary effect will be superposed on the buildup behavior and may therefore be evident in the pressure response.

Consider Example 3, a buildup test in the well shown earlier in Example 1(a) ("8.2 Bounded Reservoir Example" on page 165), if the well were to have been shut in after 21.6 hours.

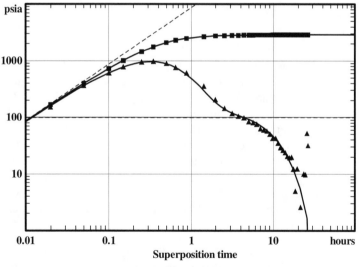

Figure 8.31

As seen earlier, no boundary effect was observed during the 21.6 hours drawdown period, although we determined from longer testing that a drainage area boundary existed at a distance of 800 feet, and that the influence of this boundary effect was significant within the 20 hours subsequent to the first 21.6 hours of drawdown.

Therefore, the boundary effect will alter the appearance of a buildup test in which the well is shut in at 21.6 hours, even though no boundary effect was visible during the drawdown. The effect of the closed boundary is to show an ever-decreasing derivative on the derivative plot, Fig. 8.31, *even when superposition time is used,* and to flatten the Horner plot towards zero slope, Fig. 8.32.

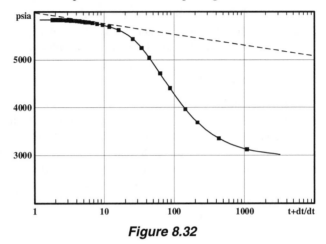

Figure 8.32

It is obviously wrong to try to diagnose this derivative plot as before, since it looks like the response of a system with a constant pressure boundary, although none is present. (Actually, remembering Example 1(a), the boundary was actually a closed circle giving rise to pseudosteady state).

The reason for this different appearance of the buildup plots is that the response consists of the summation (superposition) of the response to the change in flow rate at shut in, added to the response to the original change in flow rate when the well first started to flow. In the example shown in Figs. 8.31 and 8.32, the drawdown response has achieved pseudosteady state whereas the shut-in response is still infinite acting (with the beginning of a pseudosteady state towards the end). As a result of the summation of the two, the actual correct Horner straight line, Fig. 8.32, covers only a small portion of the data, and would probably be rather difficult to find by eye alone. Since this example was generated artificially we have been able to position the Horner straight line in Fig. 8.32 exactly based on the known values of k and s - in a real example this would not be possible.

This emphasizes a major difficulty of interpreting buildup tests, since it is not only problematic to perform reservoir diagnosis, it is also more difficult to estimate the reservoir parameters. The problems are not seen when boundary effects are not present (such as in the example illustrated in Figs. 8.28 and 8.29), and can be minimal in some cases (for example if the drawdown has reached constant pressure boundary behavior, but the buildup has not).

The difficulty of recognizing the correct Horner slope is overcome by use of nonlinear regression, however the problem of reservoir diagnosis remains. It is not usually advisable to perform diagnosis on buildup data alone, unless the response has very clearly responded as if infinite acting. Desuperposition provides a hope for diagnosis, since it separates out the constant rate response.

It has often been found in the past that the drawdown and buildup estimates of reservoir parameters were not the same. There are good physical reasons why this might occur (such as gas dissolution), however, once nonlinear regression became more commonly used it was discovered that in many cases the estimates from drawdown and buildup analyses *were* the same, and that it was the distortion of the Horner plot that caused an error in traditional estimation techniques.

8.3.3 Knowledge of p_{wf}

In the absence of wellbore storage effect, the buildup response is not affected by the skin effect, since the well is not actually flowing. Traditional analysis methods rely on the value of p_{wf}, the last pressure measured while the well was flowing, to estimate the skin factor s. If the logarithmic approximation to the infinite acting radial flow solution is used, then from Eq. 2.70:

$$s = 1.151 \left[\frac{p_{1hr} - p_{wf}}{m} - \log \frac{kt_p}{(t_p + 1)\phi\mu c_t r_w^2} + 3.2274 \right]$$

(8.3)

where the Horner slope m is 162.6 $qB\mu/kh$.

If Eq. 8.3 is used, it is seen that estimates of skin factors can be quite strongly dependent on the value of p_{wf}. This makes the traditional estimation process dependent on a single data point -- if this point is in error, then so too will be the estimate of the skin factor, s.

During nonlinear regression, skin factor is estimated differently, by matching the shape of the curve during the wellbore storage period when the well is still flowing. This estimate uses more than a single data point, and is often more consistent with estimates of skin factor obtained during drawdown, however may be different from a traditional estimate obtained using the value of p_{wf}. In a mixed graphical and computer-aided interpretation of a buildup test, it may be necessary to make a choice between forcing the regression procedure to fit the model to the p_{wf} point exactly or to fit more generally with equal weighting to all pressure data points.

As an example, the Horner plot in Fig. 8.29 provided a traditional estimate of skin factor as 6.53, using a value of p_{wf} of 2989.4 psia. The second data point, less than a minute after shut in, was nearly 200 psi higher, indicating how rapidly pressure varies at this time, and how difficult it may be to determine exactly when the well was shut. If a different value of p_{wf} is used, for example 2939.4 psi which is 50 psi different from the correct value, then the skin estimate changes to 6.86. The actual value of skin factor was 7.7, and this value is obtained by nonlinear regression, even if the wrong value of p_{wf} is used (provided the regression is not forced to pass through p_{wf} exactly).

8.4 Fractured Well Examples

As discussed in "2.10 Fractured Wells" on page 36, many wells are stimulated upon completion by hydraulic fracturing. Analysis of reservoirs that are penetrated by such wells almost always requires specific consideration of the effects of the fractures. Also, a common objective in testing fractured wells is to evaluate the effectiveness of the fracture treatment, and to estimate the length and conductivity of the fracture itself. Many hydraulically fractured wells display finite conductivity behavior, hence this type of analysis is a good way to start an interpretation if the well is known to be fractured.

8.4.1 Finite Conductivity Fracture Example

Example 4 represents a drawdown test performed when a well was put into production following a fracture stimulation. Fig. 8.33 shows the diagnostic plot, indicating the absence of wellbore storage (due to their very high productivity, hydraulically fractured wells often show very short wellbore storage effects -- this also serves the purpose of illustration in this example).

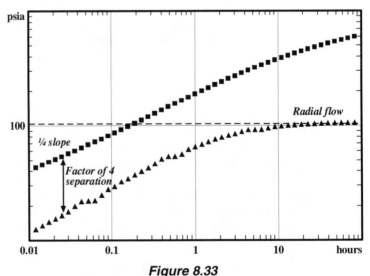

Figure 8.33

Since there is no unit slope (representing storage), since the well is known to have been fractured, and since there is the clear evidence of finite conductivity behavior (¼ slope and separation of a factor of four between the pressure and the derivative) then it is worthwhile to look at a log-log type curve plot. Fig. 8.34 shows that there is a good match to the finite conductivity type curve, with a value of $(k_f w_f/ k x_f)$ somewhere between 5 and 10.

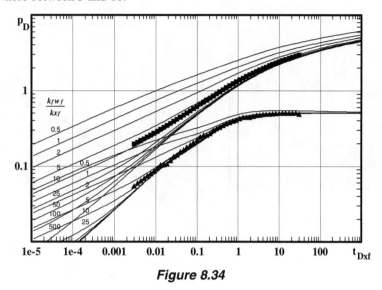

Figure 8.34

Known parameters in this example are as follows:

B_t (RB/STB)	1.5	q (STB/d)	2000
μ_o (cp)	0.3	h (feet)	50
c_t (/psi)	1.48×10^{-5}	p_i (psia)	5200
ϕ	0.24		

From the pressure match point: $\Delta p = 41.08$ psi, $p_D = 0.197$

$$p_D = \frac{kh\,\Delta p}{141.2\,qB\mu} = \frac{k(50)(41.08)}{141.2(2000)(1.5)(0.3)} = 0.197$$

Hence permeability, $k = 12.19$ md

From the time match point: $t = 0.01$ hrs, $t_{Dxf} = 2.74 \times 10^{-3}$

$$t_{Dxf} = \frac{0.000264\,k\,t}{\phi\mu c_t x_f^2} = \frac{0.000264(12.19)(0.01)}{(0.24)(0.3)(1.48 \times 10^{-5})x_f^2} = 2.74 \times 10^{-3}$$

Hence fracture length, $x_f = 105$ ft.

Based upon the pressure and time match points, the permeability and fracture length are estimated to be about 12 md and 100 ft respectively.

A plot of pressure against fourth root of time, as suggested by Eq. 2.44, shows a straight line portion with a slope of 173.3 psia/hr$^{1/4}$, with intercept at the origin corresponding to the (known) value of p_i, 5200 psia (Fig. 8.35).

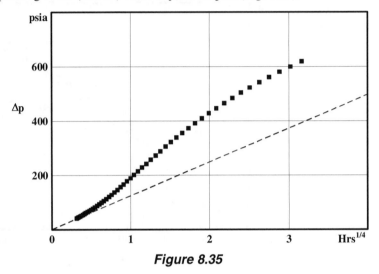

Figure 8.35

Since, from Eq. 2.44, we know that:

$$p_D = \frac{2.451}{\sqrt{k_{fD} w_{fD}}} t_{Dxf}^{1/4} \qquad (8.4)$$

then:

$$p = p_i - \frac{44.13 \, qB\mu}{h(k_f w_f)^{1/2}(\phi\mu c_t k)^{1/4}} t^{1/4} \qquad (8.5)$$

and the slope of the line p vs. $t^{1/4}$ is given by:

$$m = \frac{44.13 \, qB\mu}{h(k_f w_f)^{1/2}(\phi\mu c_t k)^{1/4}} \qquad (8.6)$$

Substituting the numbers and the slope observed in Fig. 8.35:

$$(k_f w_f)^{1/2} = -\frac{44.13(2000)(1.5)(0.3)}{(-173.3)(50)\left[(0.24)(0.3)(1.48\times10^{-5})(12.19)\right]^{1/4}}$$

Hence:

$$k_f w_f = 5829.3 \ md - ft$$

$$\frac{k_f w_f}{k x_f} = \frac{5829.3}{(12.19)(105)} = 4.55$$

This estimate of the dimensionless fracture conductivity is roughly consistent with the value 5 to 10 suggested by the type curve match.

Notice that the pressure behavior is not dependent on the length of the fracture during the bilinear flow period. The fracture length could, in principle, be estimated by locating the one half slope section of the log-log plot, however, as mentioned earlier, this half slope region is rarely present in real tests of finite conductivity fractures. The type curve match, relating values of real time t to dimensionless time t_{Dxf}, is therefore the only graphical way to estimate fracture length x_f.

Although permeability was also estimated by type curve matching in Fig. 8.34, as in unfractured reservoir analysis it is much better to calculate permeability by semilog analysis. Fig. 8.36 shows a semilog (MDH) plot, with the straight line drawn. The slope of the line provides an estimate of permeability of 12.2 md. Notice that this same line provides an estimated skin factor of -4.66.

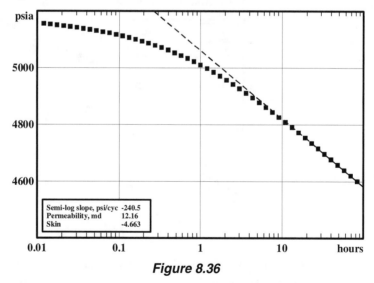

Figure 8.36

Finally, performing an automated match by nonlinear regression, the estimates are refined to $k = 12$ md, $x_f = 101$ ft, and $(k_f w_f / k \ x_f) = 9$, with good confidence intervals for all parameters, as well as a close visual match (Fig. 8.37).

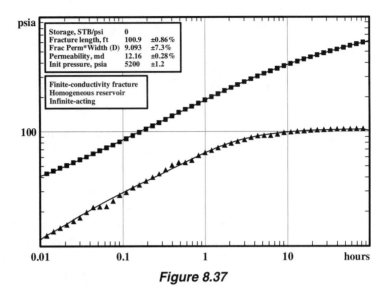

Figure 8.37

Although this looks like a straightforward example, further examination reveals some pitfalls in the analysis of finite conductivity fractures. If we match the same data to the *infinite* conductivity fracture type curves (Fig. 8.38), then the match does not look as good (although it is still quite close).

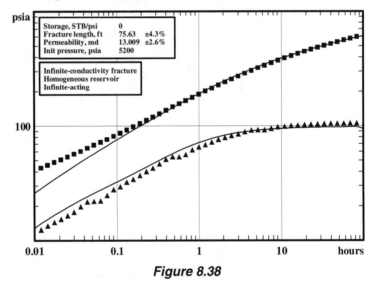

Figure 8.38

However, if we allow the initial reservoir pressure p_i to be variable, since we have already observed that this can alter the slope of the early line, then a good match to the infinite conductivity fracture model can be obtained (Fig. 8.39), by changing the value of p_i by only 20 psia.

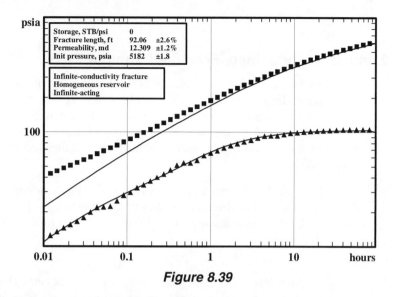

Figure 8.39

Finally, it should be noted that the presence of wellbore storage can also affect the ability to estimate fracture conductivity. The presence of even a moderate storage effect produces significant errors in the estimate of fracture conductivity (Fig. 8.40). However, the reservoir permeability is still well determined, as is the fracture half length x_f (in this case, since the storage effect does not overwhelm the entire fracture dominated portion of the response).

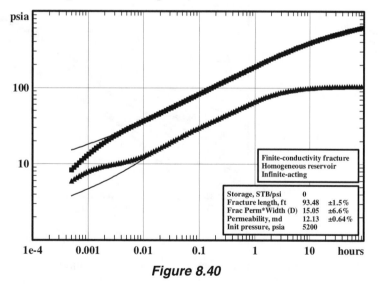

Figure 8.40

In conclusion, the interpretation of well tests in wells with finite conductivity fractures, such as those produced by hydraulic stimulation, requires careful consideration of the early data. In general, the fracture conductivity is not a very well determined parameter.

8.4.2 Infinite Conductivity Fracture Example

As can be seen from the finite conductivity fracture type curve, once the dimensionless fracture conductivity becomes large, the pressure transient behavior is indistinguishable from that of an infinite conductivity fracture, except at times that are so small that they cannot often be monitored in practice. Thus it is often simpler to treat highly conductive fractures as if they had infinite conductivity.

Example 5 represents a drawdown test in a highly productive fractured well. This test is similar to Example 4 in almost every regard, except that it can be seen that the total pressure drawdown is 35 psi less (out of a total drawdown of about 600 psi) in this case.

Looking first at the finite conductivity fracture type curve (Fig. 8.41), it is seen that the data lie very close to the high conductivity lines, and in fact show no quarter slope behavior at all. Plotting a one half slope line on the log-log graph (Fig 8.42) shows a close representation of the early data. The data also fit well to the infinite conductivity type curve (Fig. 8.43).

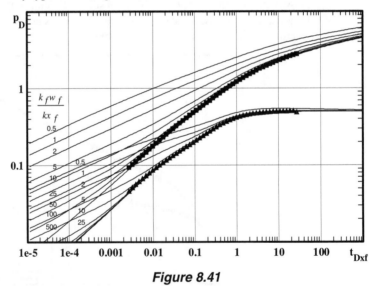

Figure 8.41

Known parameters in this example are as follows:

B_t (RB/STB)	1.5	q (STB/d)	2000
μ_o (cp)	0.3	h (feet)	50
c_t (/psi)	1.48×10^{-5}	p_i (psia)	5200
ϕ	0.24		

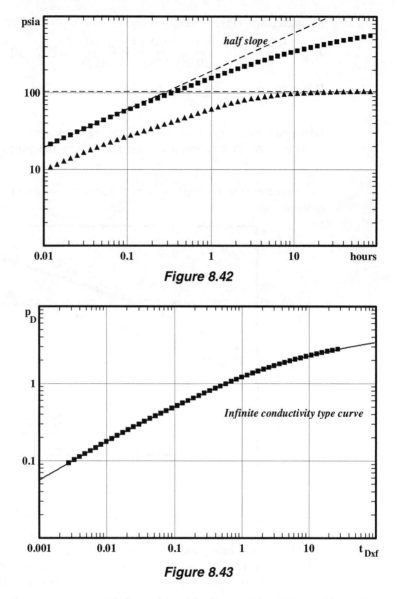

psia

half slope

100

10

0.01   0.1   1   10   hours

Figure 8.42

p_D

1

Infinite conductivity type curve

0.1

0.001   0.01   0.1   1   10 t_{Dxf}

Figure 8.43

Using the type curve match from either Fig. 8.41 or Fig. 8.43 provides estimates of k and x_f from the match points.

A match point on the pressure axis is $\Delta p = 19.48$ psi, $p_D = 9.368 \times 10^{-2}$, from which:

$$p_D = \frac{kh\,\Delta p}{141.2\,qB\mu} = \frac{k\,(50)(19.48)}{141.2\,(2000)(1.5)(0.3)} = 9.368 \times 10^{-2}$$

Hence the permeability, $k = 12.22$ md.

A match point on the time axis is $t = 0.01$ hrs, $t_{Dxf} = 2.72 \times 10^{-3}$, from which:

$$t_{Dxf} = \frac{0.000264\,k\,t}{\phi\mu c_t x_f^2} = \frac{0.000264\,(12.22)(0.01)}{(0.24)(0.3)(1.48 \times 10^{-5})x_f^2} = 2.72 \times 10^{-3}$$

Hence the fracture length, $x_f = 105.5$ ft.

Based on the match to either Fig. 8.41 or Fig. 8.43, permeability k is estimated to be 12.22 md, and fracture length x_f is estimated to be 105.5 feet.

The estimates from the type curve match are confirmed and refined by nonlinear regression (Fig. 8.44).

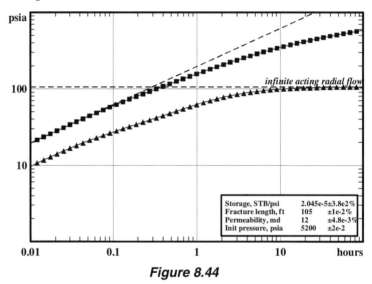

Figure 8.44

In spite of the fact that we have used only fracture type curves for this interpretation, it is important to emphasize that the permeability is still determined from the ***infinite acting radial flow*** portion of the response. This part of the data fills the last part of this example data set, and is characterized by a semilog straight line (seen in Fig. 8.45) and by a flat region in the derivative (Fig. 8.44).

In the process of drawing the semilog straight line in Fig. 8.45, it is seen that a negative skin factor of about -5 is estimated, representing a very highly stimulated well. With such a large stimulation, it is very common that the pressure transient behavior can look almost exactly like a normal storage and skin response for stimulated wells, without considering the fracture flow at all. For example, Fig. 8.46 shows a match of the example data to the storage and skin well model. The poorer match at early time can be improved somewhat by allowing the estimate of the initial pressure p_i to vary by a few psi.

Modern Well Test Analysis

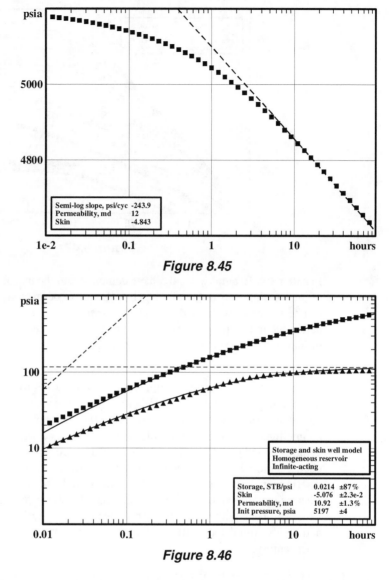

Figure 8.45

Figure 8.46

Confidence intervals shown in Fig. 8.46 reveal that all parameters except the wellbore storage coefficient were well determined.

Plotting the data on the storage and skin type curve (Fig 8.47) reveals why this so. For large negative values of skin factor s, the group $C_D e^{2s}$ takes small values, and the curves show little storage behavior in the usual form of a unit slope log-log line followed by a hump in the derivative. This may also be true for wells with finite conductivity fractures (such as that in Example 4). In fact, it is even possible to match the infinite conductivity fracture type curve itself to the storage and skin type curve.

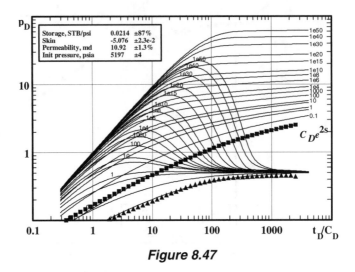

Figure 8.47

In summary, Examples 4 and 5 have demonstrated the use of type curves as a means of interpreting tests in fractured wells. Reservoir permeability and fracture length can be estimated with good accuracy. Fracture conductivity is not as easy to determine in that it gives rise only to an early time effect, and in any case does not influence the pressure transient very strongly.

The use of type curves in Examples 4 and 5 also serves to emphasize the possibility of ambiguity when doing this kind of analysis. Sometimes more than one reservoir model may fit the data, and it is necessary for the interpretation engineer to make an intelligent evaluation of all information associated with the well (such as geological and geophysical data). Nonlinear regression requires the engineer to choose a model to be matched, although confidence intervals can often be useful to discriminate between model matches. In Example 5 here, the infinite conductivity fracture model match gave a better stimulated well model match, although in this case there is no real inconsistency between the two physical models (large fracture and highly stimulated well). Both models correctly estimated the reservoir permeability.

8.5 Dual Porosity Examples

Dual porosity effects occur frequently in naturally fractured reservoirs, and may also be found in formations with thin layering sequences. The pressure transients are characterized by the presence of two separate responses, one for the "primary" porosity (the rock matrix, or lower permeability layers), and the other for the "secondary" porosity (the fractures, or high permeability layers). This has been described earlier in Section 2.11. In this section, we will examine some examples of transients in dual porosity reservoirs. Specifically, dual porosity analysis is more complex than single porosity analysis, and may require special care to take account of the effects of storage, and of producing time in the case of buildups.

8.5.1 Example with Modest Wellbore Storage

To illustrate the techniques of interpretation, Example 6 will consider a transient response in a buildup test with only modest storage effect. This is not very common in practice unless a downhole shut-off valve has been used, however serves the purpose of this example. The more realistic example, including larger storage effect, will be considered later.

Known parameters in this example are as follows:

B_t (RB/STB)	1.27	q (STB/d)	4000
μ_o (cp)	0.812	h (feet)	210
c_t (/psi)	8.3×10^{-6}	p_{wf} (psia)	5200
ϕ	0.14	t_p (hrs)	200

Without a large storage effect, the semilog (MDH) plot, Fig. 8.48, shows the expected pair of parallel semilog straight lines. (Since this test is a buildup test, we would normally use the Horner plot rather than the MDH plot for analysis, however MDH is used here since it shows the procedure more clearly).

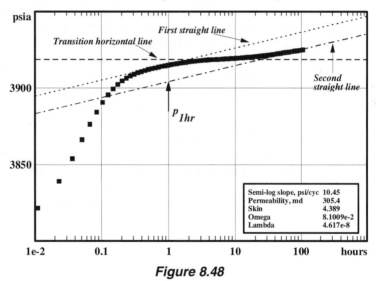

Figure 8.48

The permeability k_f of the **secondary** porosity (the fractures), can be found from the slope of either straight line, using Eq. 2.29:

$$k_f h = 162.6 \frac{qB\mu}{|m|}$$

(8.7)

Substituting values:

$$k_f(210) = 162.6 \frac{(4000)(1.27)(0.812)}{|10.45|}$$

Hence the fracture permeability, k_f = 305 md.

The skin effect s can be estimated as in a single porosity test, using the **later** semilog straight line, and substituting the value of p_{1hr} (3905 from Fig. 8.48) in Eq. 2.71:

$$s = 1.151 \left[\frac{p_{1hr} - p_{wf}}{m} - \log \frac{kt_p}{(t_p + 1)\phi\mu c_t r_w^2} + 3.2274 \right] \tag{8.8}$$

Substituting values:

$$s = 1.151 \left[\frac{3905 - 3801}{10.45} - \log \frac{(305)(200)}{(200 + 1)(0.14)(0.812)(8.3 \times 10^{-6})(0.41)^2} + 3.2274 \right]$$

Hence the skin factor, s = 5.37.

The skin effect could also be estimated from the slope of the first straight line, however must then be corrected by $+\frac{1}{2}\ln \omega$. Since ω is rather difficult to determine in many cases, and since the first straight line is often disguised by wellbore storage effects, it is usually better not use the first straight line to estimate skin unless the second straight line is not present (although in such a case, ω would be even more difficult to find).

The value of ω can be estimated from the semilog plot, by measuring the horizontal distance between the two semilog straight lines. Recall that ω reduces by a factor of ten for each log cycle between the two lines. From Fig. 8.48, it is seen that the lines are one log cycle apart, so the estimated value of ω is 0.1. To determine this estimate specifically, it is necessary only to pick two time values, t_1 and t_2, one from each line at the same pressure level. ω is then found from:

$$\omega = \frac{t_1}{t_2} \tag{8.9}$$

where t_1 is a time point on the earlier straight line, and t_2 is on the later straight line.

λ may be estimated from the level of the horizontal line passing through the dual porosity transition region. Finding the time at which the horizontal line meets the **later** semilog line (25 hrs in Fig. 8.48), λ can be found from:

$$t_D = \frac{1 - \omega}{\lambda} = \frac{0.000264 \, k_f t}{(\phi_f c_{tf} + \phi_m c_{tm})\mu r_w^2} \tag{8.10}$$

$$\frac{1-0.1}{\lambda} = \frac{0.000264\,(305)(25)}{(0.14)(8.3 \times 10^{-6})(0.812)(0.41)^2}$$

Hence $\lambda = 6.8 \times 10^{-8}$.

Examination of the derivative plot, Fig. 8.49, shows the "dip" caused by the dual porosity transition, between the two flat regions representing the two semilog straight lines. The values of ω and λ can be estimated from the position of the minimum in this dip, using a procedure outlined in "3.3.3 Estimating Dual Porosity Parameters on Derivative Plots" on page 82.

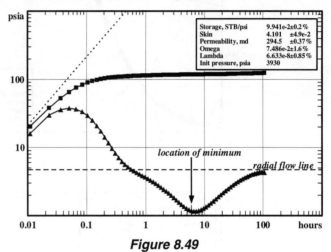

Figure 8.49

A nonlinear regression is able to obtain a good match to the data, Fig. 8.49, with acceptable confidence intervals. It is interesting to note (and is quite commonly true) that the widest confidence interval is on ω. This is commonly true with dual porosity behavior, and in fact the uncertainty is usually much greater than shown in this artificial example.

The pressure response in this example includes only a moderate storage effect ($C = 0.1$ STB/psi, duration of wellbore storage is only 0.02 hrs). It may be seen that a storage effect of longer duration could easily cover the first semilog straight line (the first "flat" portion on the derivative), although the "dip" characteristic of the dual porosity transition may still be present and recognizable. This brings us to examine the difficulties introduced into the interpretation when storage is present, as shown in the next example.

8.5.2 Example with Larger Wellbore Storage

Example 6(a) represents a modified form of Example 6, in which wellbore storage is larger. Fig. 8.50 shows the semilog (MDH) plot, which is no longer useful for interpreting the dual porosity effects since the earlier semilog straight line is not present (the lines drawn in Fig. 8.50 are in the correct position for illustration purposes only, they could not have been found from this plot). The later part of the transition is still present, allowing us to at least recognize that dual porosity effects may be present.

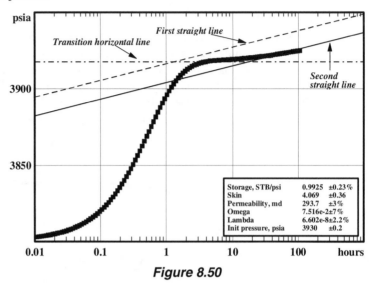

Figure 8.50

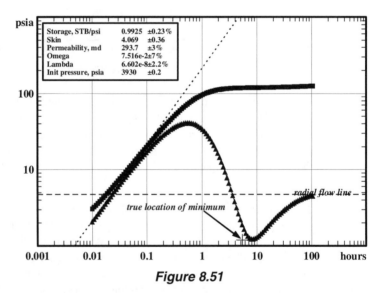

Figure 8.51

A derivative plot, Fig. 8.51, shows us the characteristic "dip". We would normally try to estimate the values of ω and λ from the position of the minimum, however caution is required here -- Fig. 8.51 shows that the position of the minimum has been displaced by the wellbore storage effect. Thus it is seen, not unexpectedly, that there is also difficulty in estimating ω and λ from the derivative plot when storage is present (we cannot expect to estimate a parameter whose influence has been hidden by the effect of another parameter, even though the derivative plot is more diagnostic in this regard than the semilog plot).

The same difficulty is found with nonlinear regression, even though this approach has the greatest precision we can achieve and can often obtain parameter estimates from transition regions. A nonlinear regression match to the data may show very poor convergence, and the final match may not look good. Luckily, the confidence intervals usually reveal that the match is unacceptable.

Just as the derivative plot shows more of the dual porosity behavior than the semilog plot, we can achieve a better nonlinear regression in dual porosity problems if we match to pressure derivatives instead of to pressures, provided the data are of sufficient quality to provide smooth derivative values. We can usually achieve a greatly improved match when derivatives or log-derivatives are used in the nonlinear regression. Recall that the confidence intervals are always wider when derivatives are used, due to the greater noise in the derivative data.

The use of pressure derivative matching by nonlinear regression is usually restricted to the dual porosity models. This is because the nonlinear regression process is as sensitive to changes in the response as is the derivative plot, and since the use of the derivative in the regression adds noise to the data. In addition, it is not possible to "float" the value of the initial pressure p_i when using the derivative for matching -- we have seen in "3.5.2 Datum Shifting" on page 87 and in "8.1.6 Pressure Shift Errors" on page 154 that it is often important to be able to do this. Thus there is little to gain by using the derivative, and often something to lose. The best approach to nonlinear regression in dual porosity analysis is to match first using log-derivatives, and then match a second time using pressures. This two step process allows a close match to be obtained first after which confidence intervals can be computed in a manner consistent with normal matching without undue influence of differentiation noise.

8.5.3 Buildup Effects

We have seen earlier in "3.3.1 Derivative Plots for Buildup Tests" on page 76, that buildup responses are more difficult to interpret, since their shape may be distorted by the transients that occurred (or *would have* occurred) in the preceding drawdown. Dual porosity tests are particularly prone to this difficulty, since they include so many different features in their response (storage, dual porosity transition, infinite acting radial flow, and boundary effect). Any of these may overlap the previous or subsequent feature depending on parameter values (for example, the response may go straight from dual porosity transition to boundary effect). In addition, the part of the response that is superposed from the drawdown transient will depend on the length of the producing time. Thus for a short producing time, the first semilog straight line in the buildup response may be superposed on the first semilog straight line in the drawdown, while for a longer producing time it may be superposed on the transition instead.

As an example of the effect of producing time t_p, Fig. 8.52 shows the Horner plot for a test, Example 6(b), with producing time 100 hours. Fig. 8.53 shows a buildup, Example 6(c), that is the same in every way except that the producing time was reduced to 25 hours. The first case ($t_p = 100$) reaches the second straight line (which is on the left in a Horner plot), while the second case ($t_p = 25$) does not. Depending on the relative durations of the drawdown and buildup time, the Horner plot may show no second straight line at all in some cases.

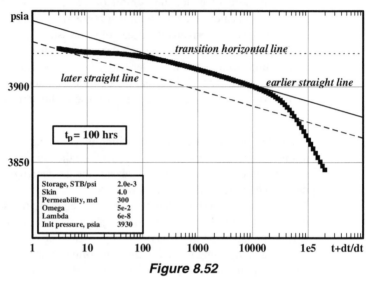

Figure 8.52

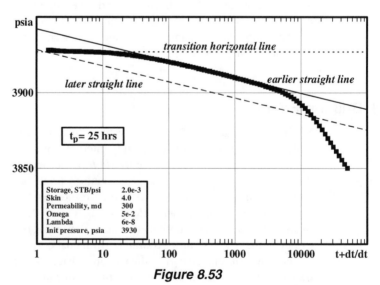

Figure 8.53

Viewing the same two examples in derivative plots, Figs. 8.54 and 8.55, it is seen that the dual porosity transition is much easier to identify than on the Horner plots.

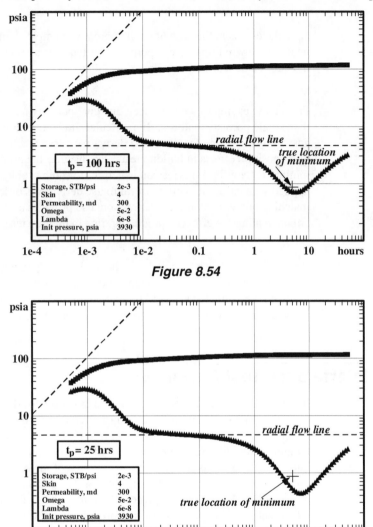

Figure 8.54

Figure 8.55

Consider the data shown in Figs. 8.53 and 8.55, with the producing time of 25 hrs. Clearly, the second straight line we would probably have drawn on the Horner plot would have given an incorrect estimate of ω and therefore also of λ. (The second straight line is shown on Fig. 8.53 in the correct position to illustrate this point). In a modern analysis, we would be more likely to estimate the values of ω and λ from the position of the minimum in the derivative, however it can be seen from Figs. 8.54 and 8.55 that the true location of the minimum may be removed from the apparent location due to superposition effects. The more severe superposition effect

---

occurs in the case when t_p is 25 hours, and this results in the greater deviation between the actual and apparent locations of the minimum.

Nonetheless, as in the case of single porosity examples, nonlinear regression is unaffected by multiple rate superpositions (such as a buildup), and is able to match the buildup data as easily as drawdown data.

From these examples we have seen that dual porosity analysis is often more difficult than single porosity analysis. In particular, it is rather easy to obtain the wrong answer by analyzing the dip in the derivative curve when storage is present. Superposition effects in buildup tests also disguise the proper location of the minimum. Nonlinear regression is helpful for both problems, however must also be applied with care -- it is often useful to use derivatives or log-derivatives in the matching process. Nonlinear regression also provides a useful way of circumventing the effect of producing time in buildup tests.

Finally, it should also be noted that the dual porosity parameters ω and λ are rather poorly estimated by any of these methods of analysis. Although we have been able to obtain the correct answers in these examples, there remains a greater uncertainty in the estimates of ω and λ than in the estimates of the other parameters.

8.6 Interference Test Examples

Interference testing differs from standard drawdown and buildup testing in that more than one well is used. One well (the "active" well) is put on production or injection, and another well (the "observation" well), which is not in production, is monitored for changes in pressure caused by operation of the first. The advantage of interference testing is that a much greater area of the reservoir is tested, providing estimates of reservoir properties between wells. In addition, the interference response is little affected by the complicating factors of wellbore storage and skin effect that make single well test interpretation more difficult. Furthermore, the nature of the response over distance makes it possible to estimate not only the reservoir transmissivity (kh), but also storativity $(\phi c_t h)$. The disadvantage is that pressure drops can be very small over distance, and are affected by other operational variations in the field at large. Nonetheless, modern electronic gauges are quite capable of registering such small pressure drops (often less than 1 psi over days or even weeks), and thus interference testing is a useful method of proving up new discoveries. In new reservoirs, an interference test is not affected by other production in the field (since there is none) and serves to prove the existence of productive reservoir between the wells.

8.6.1 Single Flow Rate Example

The following example, Example 7, is real data from a reservoir in which several wells were drilled over a wide spacing to explore the extent of the field. The producing well was 3353 feet from the observation well, and the pressure was monitored in the observation well when the production well was shut in after producing for 49 days. The test data are illustrated in Fig. 8.56. Notice that the entire pressure change in the observation well, even over 16 days, was only about 1.5 psi.

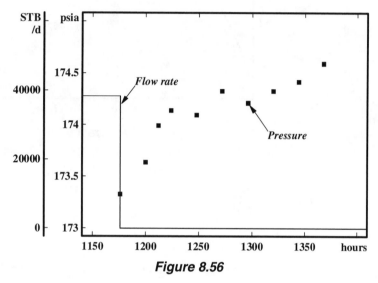

Figure 8.56

Even though an interference test still uses the same radial flow analysis as a single well test, a difference arises due to the effect of the radial distance at which the pressure is measured. The radial flow behavior in an infinite reservoir follows the exponential integral behavior:

$$p_D = -\tfrac{1}{2}\,\mathrm{Ei}\!\left(-\frac{r_D^2}{4t_D}\right)$$

which for r_D of 1 (i.e. at the wellbore) follows a behavior that is very close to log(t) (described by Eq. 2.27). However in an interference test r_D can be large, giving rise to values of t_D/r_D^2 that will be small. Over this range, the exponential integral function Ei no longer shows log(t) behavior. This is an advantage for the interpretation of interference tests, since the slope of a semilog straight line provides only one reservoir parameter estimate (kh). For small values of t_D/r_D^2 the semilog straight line develops into a curve, and it is possible to estimate both kh as well as $\phi c_t h$ by type curve matching or by nonlinear regression.

Fig. 8.57 shows a match of the data to the exponential integral type curve (sometimes known as the line source type curve, or Theis curve).

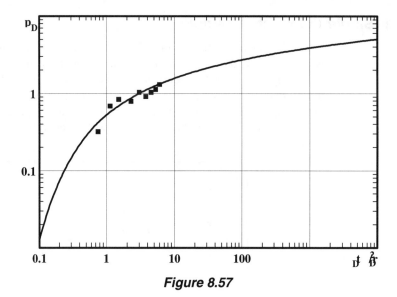

Figure 8.57

A match point in pressure is $\Delta p = 0.3087$ psi, $p_D = 0.318$.

$$p_D = \frac{kh\,\Delta p}{141.2\,qB\mu}$$

$$0.318 = \frac{kh\,(0.3087)}{141.2\,(38292)(1)(0.115)}$$

Hence the permeability-thickness product, $kh = 640400$ md-ft.

A match point in time is $t = 24$ hrs, $t_D/r_D^2 = 0.909$.

$$\frac{t_D}{r_D^2} = \frac{0.000264\,kh\,t}{\phi h \mu c_t r^2}$$

$$0.909 = \frac{0.000264\,(640400)(24)}{\phi h\,(0.115)(8.84\times10^{-6})(3353)^2}$$

Hence the porosity-thickness product, $\phi h = 390.7$ ft

The porosity-thickness product ϕh is estimated to be 390 ft. With a circular area represented by the 3353 ft radius between the wells, $(353\times10^6 \text{ ft}^2)$, the total estimated reservoir pore volume represented by these two wells would be 1.38×10^{10} ft^3. This is a very valuable parameter to have been able to estimate, as it bears directly on the reserve estimates for the field.

Nonlinear regression leads to similar estimates. In this example the confidence intervals are very wide, mainly due to the scarcity of data as well as to the relative noise.

To check the consistency of the estimates, we could also look at the semilog (MDH or Horner) plot, since the later part of the data may reach to a semilog behavior (it need not necessarily always do so). The intercept of the semilog line should indicate a skin effect of zero, as must be the case if the interpretation is correct in using the radial flow model.

8.6.2 Multirate Example

From Example 7 it has been seen that an interference test provides very valuable information during the evaluation of a reservoir, but that interpretation may be made more difficult since noise tends to mask the very small pressure changes. Considerable improvement to the interpretation can be made if more than one flow period is included in the data.

Example 8 represents an interference test in which water was injected into the active well, and the pressure monitored in the observation well both during the injection as well as during the subsequent falloff. This same example appears in the monograph by Earlougher (1977) as his Example 9.1. The continuation of monitoring after shut-in provides a longer set of data, and also provides a consistency check. Fig. 8.58 shows the flow rate in the active well, and the corresponding response in the observation well.

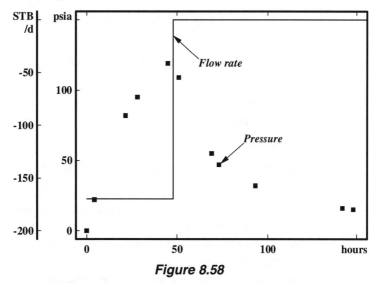

Figure 8.58

This data can be matched to a special form of the exponential integral or line source type curve developed by Henry J. Ramey, Jr., of Stanford University in 1981. This type curve (Fig. 8.59) includes both drawdown and buildup periods on the same type curve, and can be used to match both parts of the data at the same time.

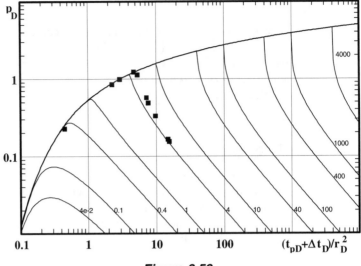

Figure 8.59

A match point in pressure is $\Delta p = 22$ psi, $p_D = 0.227$.

$$p_D = \frac{kh \, \Delta p}{141.2 \, qB\mu}$$

$$0.227 = \frac{kh\,(0.22)}{141.2\,(170)(1)(1)}$$

Hence the permeability-thickness product, $kh = 247.86$ md-ft.

A match point in time is $t = 43.7$ hrs, $t_D/r_D^2 = 4.558$.

$$\frac{t_D}{r_D^2} = \frac{0.000264 \, kh \, t}{\phi h \mu c_t r^2}$$

$$4.558 = \frac{0.000264\,(247.86)(43.7)}{\phi h\,(1)(9 \times 10^{-6})(119)^2}$$

Hence the porosity-thickness product, $\phi h = 4.922$ ft.

Nonlinear regression achieves a good match to both parts of the response at the same time (since nonlinear regression can accommodate any sequence of flow rates), and shows a good visual match (Fig 8.60) as well as acceptable confidence intervals (Table 8.9).

Parameter	Estimate	Confidence Interval (absolute)	Confidence Interval (percent)
kh (md-ft)	247.8	±14.21	±5.73%
ϕh (ft)	4.922	±0.4936	±10.03%

Table 8.9

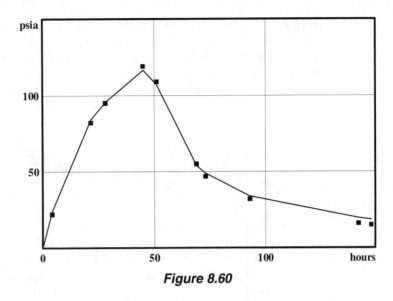

Figure 8.60

8.7 Gas Well Test Example

Example 9 represents a flow-after-flow drawdown test in which a gas well was produced at five successively greater flow rates for four hours each. Figure 8.61 shows the pressure and flow rate as functions of time.

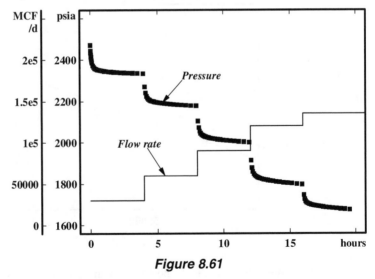

Figure 8.61

Analyzing the semilog plots is made more difficult by the need to specify a flow rate, while there were five different flow rates during the test. A multirate (rate normalization) plot, as in Fig. 8.62, illustrates the straight line portions of the data suggestive of infinite acting radial flow. However, the skin effect is seen to progressively increase with each increase in flow rate (compare the skin factor estimate of 3.262 from the first flow rate line in Fig. 8.62 with the estimate of 4.008 from the third flow rate line). It is clear that there is a rate dependent skin effect in this case. The rate dependent skin effect can be estimated by making a plot of apparent skin effect as a function of flow rate, as in Fig. 8.63.

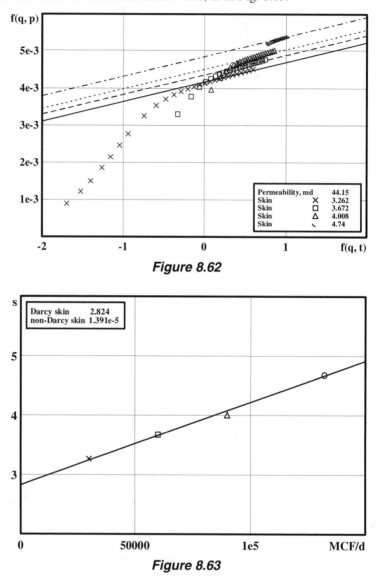

Figure 8.62

Figure 8.63

Nonlinear regression is the best way to analyze this data set, since the constant and rate dependent skin effects can be estimated from the consideration of all five flow rates together. Table 8.10 lists the parameter estimates and the confidence intervals, and Fig. 8.64 shows the match.

Parameter	Estimate	Confidence Interval (absolute)	Confidence Interval (percent)
k (md)	34.81	±0.418	±1.2%
s	1.11	±0.23	±21%
D (/MCF/d)	9.95×10^{-6}	$\pm 3.58 \times 10^{-7}$	±3.6%
C (SCF/psi)	3.225	±0.103	±3.2%

Table 8.10

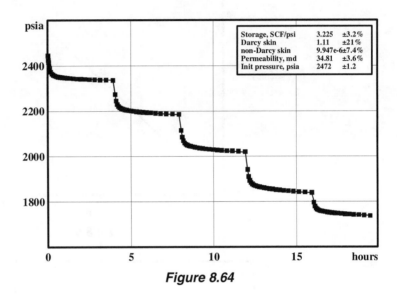

Figure 8.64

Once skin factor s and rate dependent skin factor D have been estimated in this way, the open flow potential ($AOFP$) can be estimated by simulating a multirate test down to the minimum achievable well head pressure. The fact that pressures may not have stabilized during the actual test is of no consequence, since the simulated test can have individual flow periods as long as necessary to achieve stabilization.

Parameter	Estimate	Confidence Interval (absolute)	Confidence Interval (percent)
k (md)	35.61	±2.24	±6.3%
s	-0.17	±0.03	±19%
D (/MCF/d)	14.73×10^{-6}	$\pm 1.47 \times 10^{-6}$	±10%
C (SCF/psi)	2.993	±0.176	±5.9%

Table 8.11

The use of pseudopressure is not always necessary for proper interpretation. Fig. 8.65 shows a plot of pseudopressure against pressure for the data of this test. Since the relationship is almost completely linear, there would be little consequence if the analysis had used pressure directly. Fig. 8.66 shows a nonlinear regression match using pressure and Table 8.11 shows the parameter estimates.

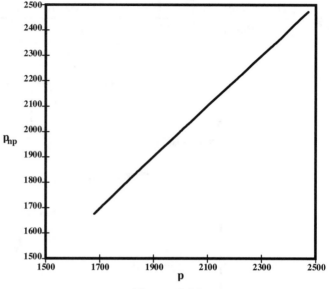

Figure 8.65

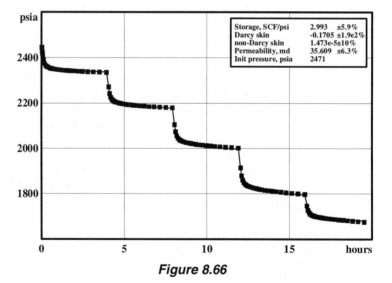

Storage, SCF/psi	2.993	±5.9%
Darcy skin	-0.1705	±1.9e2%
non-Darcy skin	1.473e-5	±10%
Permeability, md	35.609	±6.3%
Init pressure, psia	2471	

Figure 8.66

8.8 References

Earlougher, R.C., Jr.: *"Advances in Well Test Analysis"*, Society of Petroleum Engineers Monograph 5, Dallas, TX, (1977).

9. CALCULATING PROPERTIES

9.1 Introduction

In order to use most of the methods described here, it is necessary to first obtain numerical values for fluid and rock properties such as formation volume factor B and total system compressibility c_t. Sometimes these parameters have been measured directly in the laboratory by core or PVT analysis, but often it is necessary to estimate them. This section discusses methods of estimation, so that the required oil, gas, water and rock properties can be determined if they are not otherwise available. These estimation techniques are based upon observations and correlations developed by many authors. Many of the correlations described here have been collected in a book by McCoy (1983).

9.2 Oil Properties

The oil properties required for most well test applications are formation volume factor B_o, viscosity μ_o, and compressibility c_o (which is included into the calculation of total compressibility c_t). These parameters are functions of the reservoir pressure and temperature, the separator pressure and temperature, the gas/oil ratio and gas gravity, the API density of the oil, and depend on whether the reservoir pressure is above or below the bubble point.

Following the sequence of calculations described by McCoy (1983), the first step is to correct the observed gas gravity to separator conditions:

$$\gamma_{gs} = \gamma_g \left[1 + 5.912 \times 10^{-5} \; API \; T_s \; \log\left(\frac{p_s}{114.7}\right) \right]$$

(9.1)

where:

γ_{gs}	=	gas gravity corrected to separator conditions
γ_g	=	observed gas gravity (air = 1)
API	=	observed oil API gravity
T_s	=	separator temperature (°F)
p_s	=	separator pressure (psia)

The second step is to calculate the bubble point pressure p_{bp} in psia, based upon the observed gas oil ratio GOR in SCF/STB, and the reservoir temperature T_R in °R:

$$For \quad API \leq 30, \quad p_{bp} = \left[\frac{GOR}{0.0362\,\gamma_{gs}e^{\frac{25.724\,API}{T_R}}} \right]^{0.9143}$$

(9.2)

$$For \quad API > 30, \quad p_{bp} = \left[\frac{GOR}{0.0178\,\gamma_{gs}e^{\frac{23.931\,API}{T_R}}} \right]^{0.8425}$$

(9.3)

The third step is to calculate the equilibrium gas/oil ratio R_s in SCF/STB:

$$For \quad API \leq 30, \quad R_s = 0.0362\,\gamma_{gs}\,p^{1.0937}\,e^{\frac{25.724\,API}{T_R}}$$

(9.4)

$$For \quad API > 30, \quad R_s = 0.0178\,\gamma_{gs}\,p^{1.1870}\,e^{\frac{23.931\,API}{T_R}}$$

(9.5)

In calculations that require a value of the reservoir **dissolved** gas/oil ratio GOR_{res}, the value of observed GOR is used if reservoir pressure is greater than bubble point pressure ($p > p_{bp}$). If reservoir pressure is less than bubble point ($p < p_{bp}$), then the equilibrium gas/oil ratio R_s is used for GOR_{res}.

The fourth step is to calculate the oil compressibility c_o (/psi) at or above the bubble point:

$$c_o = \frac{-2433 + 5\,GOR + 17.2\,T_F - 1180\,\gamma_{gs} + 12.61\,API}{10^5\,p}$$

(9.6)

where:

T_F = reservoir temperature (°F)

p = reservoir pressure (psia)

The fifth step is to calculate the oil viscosity. Different equations are used depending on whether the oil is above or below the bubble point. It is first necessary to calculate viscosity of the "dead oil" (gas free oil) μ_{od} in cp:

$$\mu_{od} = 10^x - 1$$

(9.7)

where $x = 10^{3.0324 - 0.02023\,API}\,T_F^{-1.163}$ (9.8)

After which the viscosity of the "live oil" at or below the bubble point μ_{ob} in cp is given by:

$$\mu_{ob} = 10.715(GOR_{res} + 100)^{-0.515}\,\mu_{od}^{x}$$ (9.9)

where $x = 5.44\,(GOR_{res} + 150)^{-0.338}$ (9.10)

The viscosity of the "live oil" above the bubble point μ_o in cp is given by:

$$\mu_o = \mu_{ob}\left(\frac{p}{p_{bp}}\right)^{x}$$ (9.11)

where $x = 2.6\,p^{1.187}\,e^{-0.0000898\,p - 11.513}$ (9.12)

Finally the oil formation volume factor is calculated. At or below the bubble point, B_{ob} is given by:

$$For \quad API \le 30, \quad B_{ob} = 1 + 1.751 \times 10^{-5}(T_F - 60)\left(\frac{API}{\gamma_{gs}}\right)$$

$$+ \left[4.677 \times 10^{-4} - 1.811 \times 10^{-8}(T_F - 60)\left(\frac{API}{\gamma_{gs}}\right)\right]GOR_{res}$$ (9.13)

$$For \quad API > 30, \quad B_{ob} = 1 + 1.100 \times 10^{-5}(T_F - 60)\left(\frac{API}{\gamma_{gs}}\right)$$

$$+ \left[4.670 \times 10^{-4} + 1.377 \times 10^{-9}(T_F - 60)\left(\frac{API}{\gamma_{gs}}\right)\right]GOR_{res}$$ (9.14)

The oil formation volume factor above the bubble point B_o in RB/STB is given by:

$$B_o = B_{ob}\,e^{c_o(p_{bp} - p)}$$ (9.15)

9.3 Gas Properties

The gas properties required for most well test applications are formation volume factor B_g, viscosity μ_g, and compressibility c_g (which is included into the calculation of total compressibility c_t). These parameters are functions of the reservoir pressure and temperature, the critical (or pseudocritical) pressure and temperature, the gas gravity, and the concentrations of the gas impurities N_2, CO_2 and H_2S.

The first step is to determine the pseudocritical pressure p_{pc} in psia and temperature T_{pc} in °R from the gas gravity γ_g, using Standing's correlation:

For California gases:

$$p_{pc} = 677 + 15\gamma_g - 37.5\gamma_g^2 \tag{9.16}$$

$$T_{pc} = 168 + 325\gamma_g - 12.5\gamma_g^2 \tag{9.17}$$

For condensate gases:

$$p_{pc} = 706 - 51.7\gamma_g - 11.1\gamma_g^2 \tag{9.18}$$

$$T_{pc} = 187 + 330\gamma_g - 71.5\gamma_g^2 \tag{9.19}$$

If unknown, gas gravity can also be calculated from measured values of critical pressure p_c in psia and temperature T_c in °R if these are available:

$$\gamma_g = \frac{1}{2}\left[\frac{T_c - 175.59}{307.97} - \frac{p_c - 700.55}{47.94}\right] \tag{9.20}$$

If the gas contains impurities, corrections should be made using the correction e in °F as described by Wichert and Aziz (1972):

$$e = 120(y_{CO_2} + y_{H_2S})^{0.9} - 120(y_{CO_2} + y_{H_2S})^{1.6} + 15(y_{H_2S}^{0.5} - y_{H_2S}^4) \tag{9.21}$$

$$p_c^* = \frac{p_c(T_c - e)}{T_c + y_{H_2S}(1 - y_{H_2S})e} \tag{9.22}$$

$$T_c^* = T_c - e \tag{9.23}$$

where:

y_{CO_2} = CO_2 content (decimal)

y_{H_2S} = H_2S content (decimal)

T_c^* = corrected critical temperature (°R)

p_c^* = corrected critical pressure (psia)

Having estimated or obtained the pseudocritical or critical properties, it is necessary to calculate the reduced pressure and temperature, p_r and T_r, as follows:

$$p_r = \frac{p}{p_{pc}}$$

(9.24)

$$T_r = \frac{T}{T_{pc}}$$

(9.25)

The reduced density ρ is calculated iteratively using Newton's method, and used to estimate the z factor with the procedure described by Dranchuk, Purvis and Robinson (1974) to evaluate the Standing and Katz (1942) relations. The $(k+1)$th estimate is determined from the kth estimate using the equation:

$$\rho_{k+1} = \rho_k - \frac{f(\rho_k)}{f'(\rho_k)}$$

(9.26)

where:

$$f(\rho) = a\rho^6 + b\rho^3 + c\rho^2 + d\rho + e\rho^3(1 + f\rho^2)\exp[-f\rho^2] - g$$

(9.27)

$$f'(\rho) = 6a\rho^5 + 3b\rho^2 + 2c\rho + d + e\rho^2(3 + f\rho^2[3 - 2f\rho^2])\exp(-f\rho^2)$$

(9.28)

where:

$a = 0.06423$

$b = 0.5353\,T_r - 0.6123$

$c = 0.3151\,T_r - 1.0467 - 0.5783 / T_r^2$

$d = T_r$

$e = 0.6816 / T_r^2$

$f = 0.6845$

$g = 0.27\,p_r$

$\rho_0 = 0.27\,p_r / T_r$

Then $z = \dfrac{0.27\,p_r}{\rho T_r}$

(9.29)

The gas compressibility is calculated next, from the expression:

$$\frac{1}{c_g} = p_c p_r \left[1 + \frac{\rho}{z}\frac{\partial z}{\partial \rho} \right]$$

(9.30)

where:

$$\frac{\partial z}{\partial \rho} = \frac{1}{\rho T_r}\left[5a\rho^5 + 2b\rho^2 + c\rho + 2e\rho^2\left(1 + f\rho^2 - f^2\rho^4\right)\exp\left(-f\rho^2\right)\right]$$

(9.31)

and where a, b, c, d, e, f are as in Eqs. 9.27 and 9.28

The next step is the calculation of gas viscosity, μ_g. This is done in two stages, first the Carr, Kobayashi and Burrows (1954) gas viscosity μ_{g1} is determined, after which the equation of Dempsey (1965) is used for the final result.

$$\mu_{g1} = \left(1.709 \times 10^{-5} - 2.062 \times 10^{-6}\gamma_g\right)T_F + 8.188 \times 10^{-3} - 6.15 \times 10^{-3}\log\gamma_g$$

$$+ y_{N_2}\left(8.48 \times 10^{-3}\log\gamma_g + 9.59 \times 10^{-3}\right)$$

$$+ y_{CO_2}\left(9.08 \times 10^{-3}\log\gamma_g + 6.24 \times 10^{-3}\right)$$

$$+ y_{H_2S}\left(8.49 \times 10^{-3}\log\gamma_g + 3.73 \times 10^{-3}\right)$$

(9.32)

$$\ln\left(T_r\frac{\mu_g}{\mu_{g1}}\right) = a_0 + a_1 p_r + a_2 p_r^2 + a_3 p_r^3$$

$$+ T_r\left(a_4 + a_5 p_r + a_6 p_r^2 + a_7 p_r^2\right)$$

$$+ T_r^2\left(a_8 + a_9 p_r + a_{10} p_r^2 + a_{11} p_r^2\right)$$

$$+ T_r^3\left(a_{12} + a_{13} p_r + a_{14} p_r^2 + a_{15} p_r^2\right)$$

(9.33)

$a_0 = -2.46211820$

$a_1 = 2.97054714$

$a_2 = -2.86264054 \times 10^{-1}$

$a_3 = 8.05420522 \times 10^{-3}$

$a_4 = 2.80860949$

$a_5 = -3.49803305$

$a_6 = 3.60373020 \times 10^{-1}$

$a_7 = -1.04432413 \times 10^{-2}$

$a_8 = -7.93385684 \times 10^{-1}$

$a_9 = 1.39643306$

$a_{10} = -1.49144925 \times 10^{-1}$

$a_{11} = 4.41015512 \times 10^{-3}$

$$a_{12} = 8.39387176 \times 10^{-2}$$

$$a_{13} = -1.86408848 \times 10^{-1}$$

$$a_{14} = 2.03367881 \times 10^{-2}$$

$$a_{15} = -6.09579263 \times 10^{-4}$$

Finally, the gas formation volume factor B_g (reservoir ft^3/SCF) is calculated using reservoir pressure p, reservoir temperature T_R in °R, and the z factor calculated previously:

$$B_g = 0.02829 \frac{zT_r}{p}$$

(9.34)

9.4 Water Properties

The water properties required for most well test applications are formation volume factor B_w, viscosity μ_w, and compressibility c_w (which is included into the calculation of total compressibility c_t). These parameters are functions of the reservoir pressure and temperature, as well as the salinity of the water.

Starting first with the water compressibility c_w (/psi):

$$c_{w1} = \left(a + bT_F + cT_F^2\right)10^{-6}$$

(9.35)

where:

$$a = 3.8546 - 0.000134\,p$$

$$b = -0.01052 + 4.77 \times 10^{-7}\,p$$

$$c = 3.9267 \times 10^{-5} - 8.8 \times 10^{-10}\,p$$

Correcting for the salinity $NaCl$ (in percent, 1% = 10,000 ppm):

$$c_w = c_{w1}\left[1 + NaCl^{0.7}\left(-0.052 + 0.00027\,T_F - 1.14 \times 10^{-6}\,T_F^2 + 1.121 \times 10^{-9}\,T_F^3\right)\right]$$

(9.36)

The water formation volume factor, B_w in RB/STB, is determined from:

$$B_w = \left(a + b\,p + c\,p^2\right)S_{c1}$$

(9.37)

where for gas-free water:

$$a = 0.9947 + 5.8 \times 10^{-6} T_F + 1.02 \times 10^{-6} T_F^2$$

$$b = -4.228 \times 10^{-6} + 1.8376 \times 10^{-8} T_F - 6.77 \times 10^{-11} T_F^2$$

$$c = 1.3 \times 10^{-10} - 1.3855 \times 10^{-12} T_F + 4.285 \times 10^{-15} T_F^2$$

and for gas-saturated water:

$$a = 0.9911 + 6.35 \times 10^{-6} T_F + 8.5 \times 10^{-7} T_F^2$$

$$b = -1.093 \times 10^{-6} - 3.497 \times 10^{-9} T_F + 4.57 \times 10^{-12} T_F^2$$

$$c = -5 \times 10^{-11} + 6.429 \times 10^{-13} T_F - 1.43 \times 10^{-15} T_F^2$$

and where the salinity correction factor S_{c1} is given by:

$$S_{c1} = 1 + NaCl \left[5.1 \times 10^{-8} p + (5.47 \times 10^{-6} - 1.96 \times 10^{-10} p)(T_F - 60) \right.$$
$$\left. + (-3.23 \times 10^{-8} + 8.5 \times 10^{-13} p)(T_F - 60)^2 \right]$$

(9.38)

The water viscosity μ_w in cp is given by:

$$\mu_w = S_{c2} \, S_p \, 0.02414 \times 10^{446.04/(T_R - 252)}$$

(9.39)

where the salinity correction, S_{c2}, and pressure correction, S_p, are given respectively by:

$$S_{c2} = 1 - 0.00187 \, NaCl^{0.5} + 0.000218 \, NaCl^{2.5}$$
$$+ \left(T_F^{0.5} - 0.0135 \, T_F \right) \left(0.00276 \, NaCl - 0.000344 \, NaCl^{1.5} \right)$$

(9.40)

$$S_p = 1 + 3.5 \times 10^{-12} p^2 (T_F - 40)$$

(9.41)

9.5 Rock Properties

The rock properties required for most well test applications are porosity ϕ and compressibility c_r (which is included into the calculation of total compressibility c_t). In general, it is better to use direct measurements of these properties, since there are no widely accepted correlations for them. Since experimental observations have wide scatter, there is little expectation that measurements from one reservoir will necessarily apply to another reservoir (or even a different part of the same reservoir).

However, in cases where compressibility measurements are unavailable, c_r in units of /psi can be estimated from experimental observations summarized by Newman (1973). These observations are provided here in the form of correlations, although there was considerable scatter in the original data. It is important to note that porosity in these correlations is given as a decimal number between 0.0 and 1.0.

For consolidated limestones:

$$c_r = \exp\left(4.026 - 23.07\,\phi + 44.28\,\phi^2\right) \times 10^{-6}$$

(9.42)

For consolidated sandstones:

$$c_r = \exp\left(5.118 - 36.26\,\phi + 63.98\,\phi^2\right) \times 10^{-6}$$

(9.43)

For unconsolidated sandstones:

$$For\ \phi \geq 0.2,\quad c_r = \exp\left(34.012[\phi - 0.2]\right) \times 10^{-6}$$

(9.44)

These three correlations are illustrated in Fig. 9.1.

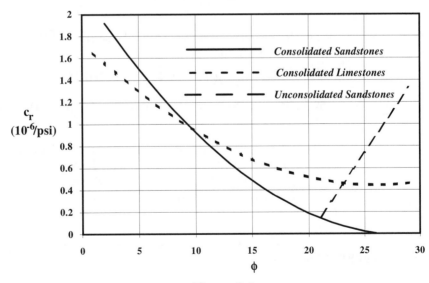

Figure 9.1

9.6 Total Properties

The final calculation is of the total compressibility c_t. For single phase flow, c_t is given by:

$$c_t = c_r + S_o c_o + S_w c_w + S_g c_g \qquad (9.45)$$

where S_o, S_w and S_g are the reservoir oil, water and gas saturations respectively.

For multiphase flow in the reservoir, c_t is given by Eq. 5.7, repeated here:

$$c_t = c_r + S_o c_o + S_w c_w + S_g c_g + \frac{S_o B_g}{5.615\,B_o}\left(\frac{\partial R_s}{\partial p}\right) + \frac{S_w B_g}{5.615\,B_w}\left(\frac{\partial R_{sw}}{\partial p}\right) \qquad (9.46)$$

It is important to consider the formation volume factor B, which appears always in the product qB throughout all the well test analysis methods that have been described here. The product qB represents the impulse imposed on the reservoir to which the measured pressure transient is the response. Therefore qB is the total voidage from the reservoir, in reservoir volumes per unit time. Thus the choice of which formation volume factor B to use depends on which q is being specified. In most cases the flow rate specified is the **oil** flow rate q_o, therefore the correct B to use would be the **total** formation volume factor B_t.

The total formation volume factor B_t can be determined from the component formation volume factors:

$$B_t = B_o + B_w\,WOR + B_g\left(1000\,GOR - R_s - R_{sw}\,WOR\right)/\,5.615 \qquad (9.47)$$

where WOR is the producing water-oil ratio and GOR is the producing gas-oil ratio.

Finally, it is worth noting that it is usual to use the **drilled radius** for the wellbore radius r_w. The actual effective wellbore radius depends considerably on the type of well completion and on the condition of the well, however the use of drilled hole radius provides a consistent way of specifying r_w.

9.7 References

Carr, N.L., Kobayashi, R., and Burrows, D.B.: "Viscosity of Hydrocarbon Gases Under Pressure", *Trans.*, AIME, (1954), 263-272.

Dempsey, J.R.: "Computer Routine Treats Gas Viscosity as a Variable", *Oil & Gas Journal*, (August 1965), 141-145.

Dranchuk, P.M., Purvis, R.A., and Robinson, D.B.: *"Computer Calculations of Natural Gas Compressibility Factors Using the Standing and Katz Correlation"*, Institute of Petroleum Technical Series, No. IP-74-008, (1974).

McCoy, R.L.: *"Microcomputer Programs for Petroleum Engineers: 1. Reservoir Engineering and Formation Evaluation"*, Gulf Publishing Co., Houston, (1983).

Newman, G.H.: "Pore-Volume Compressibility of Consolidated, Friable, and Unconsolidated Rocks Under Hydrostatic Loading," *J. Pet. Tech.*, (1973), 129-134.

Standing, M.B., and Katz, D.L.: "Density of Natural Gases", *Trans.*, AIME, (1942), 140-144.

Wichert, E., and Aziz, K.: "Calculating Z's for Sour Gases", *Hydrocarbon Processing*, (May 1972), 51-55.

10. FIELD EXAMPLES

Test 1: Stimulated Well

This buildup test followed a year or more of production. Since the exact length of the flowing period is unknown (but is known to be long), the test is treated as a drawdown, with the flow rate change set to be negative. This well has been stimulated by acidization.

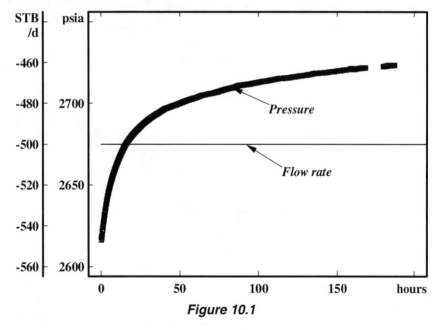

Figure 10.1

The diagnostic plot shows a typical behavior for a stimulated well; the lack of a hump in the derivative is indicative of negative skin. The appearance of the late time derivative is noise -- this is common and should not be confused with a boundary effect. Confidence intervals are excellent for all variables. The data can also be matched to a fractured well model with short fracture length (Fig. 10.3), again with good confidence.

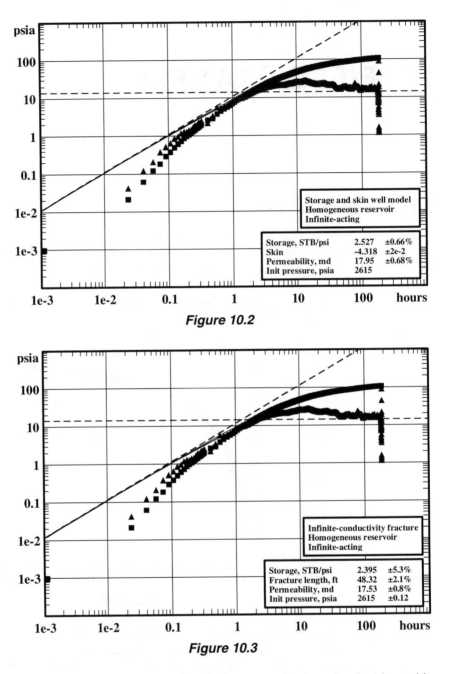

Figure 10.2

Figure 10.3

The appearance of the early time data (in the storage dominated region) is sensitive to the value of initial pressure (p_i). Even 0.1 psi change in the value of the initial pressure will influence appearance of the early time behavior. The measured data at early time do not follow a unit slope. The reason is that the time scale in the data is slightly misaligned -- this is evident from the fact that both the pressure and the derivative are in error. Subtracting 0.064 hours (five minutes) from the time scale results in a much better looking plot (Fig. 10.4).

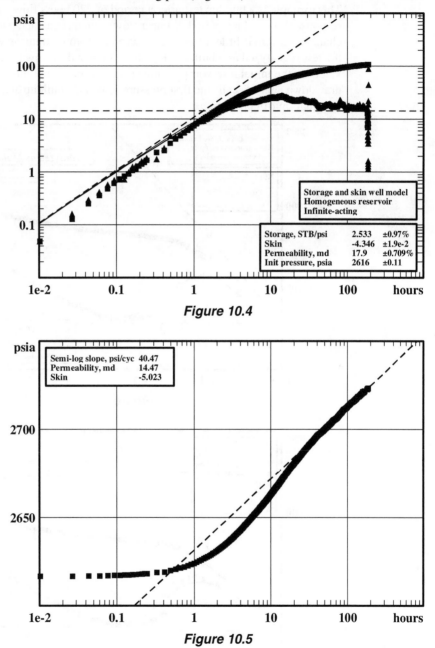

Figure 10.4

Figure 10.5

A conventional graphical (semilog) analysis as shown in Fig. 10.5 would probably underestimate k and s. This is because of the human tendency to want to draw the semilog straight line back further down the data than is appropriate.

Test 2: Channel Sand

This example is a buildup test after a period of 720 hours flow. The late time response shows a long period of linear flow, which is characteristic of flow in channels. There is little difference between reservoir parameter estimates whether the test is analyzed as a buildup (Fig. 10.6) or as a drawdown (Fig. 10.7). The values of the initial reservoir pressure appear different, but this is only because a drawdown analysis calls the first pressure p_i while a buildup analysis calls the eventual shut-in pressure p_i.

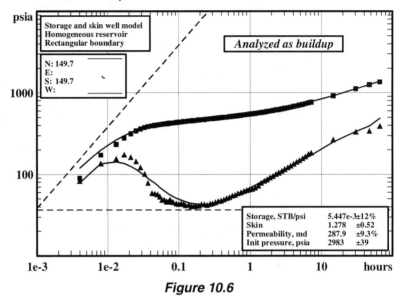

Figure 10.6

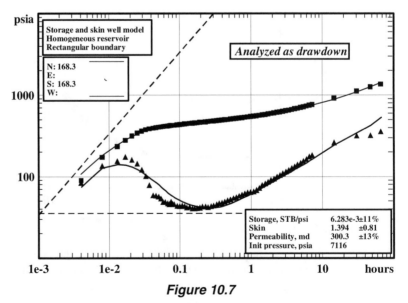

Figure 10.7

Test 3: Closed Drainage Area

This is a buildup test in a gas reservoir. Unlike Test 2, this example shows a major difference when analyzed as a drawdown (Fig. 10.8) or as a buildup (Fig. 10.9). As a "drawdown", the unit slope at late time could be matched as a pseudosteady state response (closed drainage region) (Fig. 10.8). But treating the response as a buildup cannot match the late time behavior (Fig. 10.9) if we try to impose a pseudosteady state model, the derivative dips down instead of up late time.

The example is actually a buildup test, so it may seem surprising that the late time behavior shows an increasing derivative. We know that the derivative must always go down for a buildup test, because the pressure stabilizes. However in the case of an <u>open pattern</u> (such as a channel, which we saw in Test 2) the derivative does go up, although it must eventually turn around and come down again. This test can be matched with a three-sided rectangle or "U-shaped" boundary (Fig. 10.10)

This example has been analyzed using normalized pseudopressure. Since the pressures are high (about 6000 psia), it is not necessary to use pseudotime.

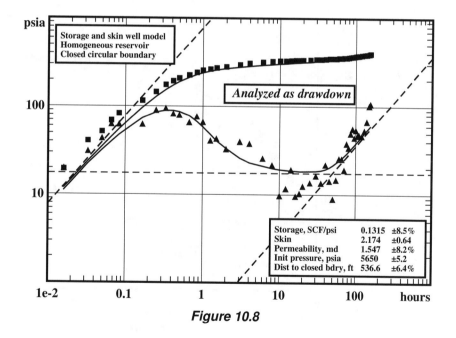

Figure 10.8

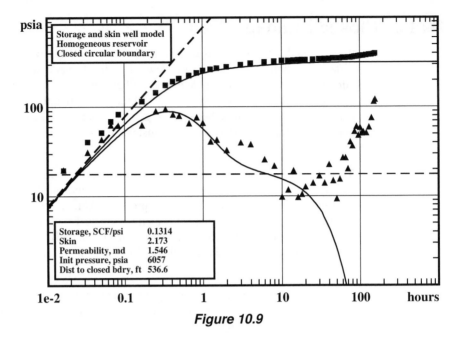

Storage and skin well model
Homogeneous reservoir
Closed circular boundary

Storage, SCF/psi	0.1314
Skin	2.173
Permeability, md	1.546
Init pressure, psia	6057
Dist to closed bdry, ft	536.6

Figure 10.9

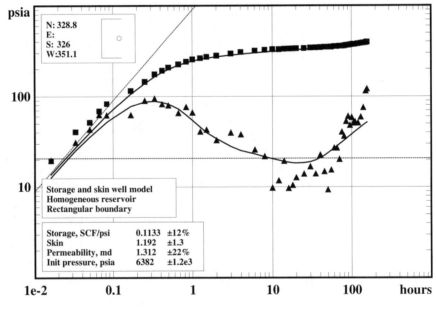

N: 328.8
E:
S: 326
W:351.1

Storage and skin well model
Homogeneous reservoir
Rectangular boundary

Storage, SCF/psi	0.1133	±12%
Skin	1.192	±1.3
Permeability, md	1.312	±22%
Init pressure, psia	6382	±1.2e3

Figure 10.10

Test 4: Wedge Shaped Boundary

This buildup test shows an unusual 1/3 slope at late time. It can be matched to a
60° wedge boundary model (Fig. 10.11). Confidence interval on permeability is
poor, since there is no infinite-acting radial flow period (the response goes straight
from storage to boundary effect). Experimenting with adding dual porosity analysis
significantly improves the match (Fig. 10.12), with much better confidence
intervals.

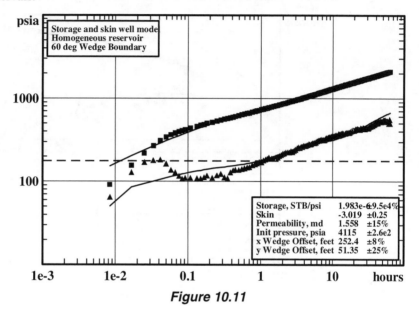

Figure 10.11

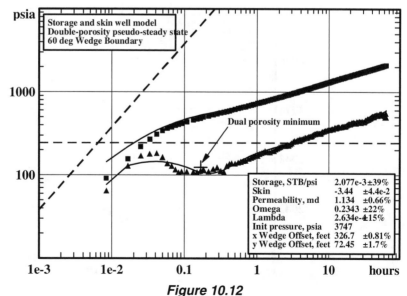

Figure 10.12

Test 5: Transition Data

This is a buildup test that has 3 hours of shut-in data following 18 hours of production. All of the data lie in the transition between storage and radial flow, however the reservoir parameters can be estimated with good confidence intervals using nonlinear regression (Fig. 10.13). Traditional Horner analysis tends to underestimate both permeability and skin -- this is generally true, but is exacerbated in this example because the data are so short (Fig. 10.14).

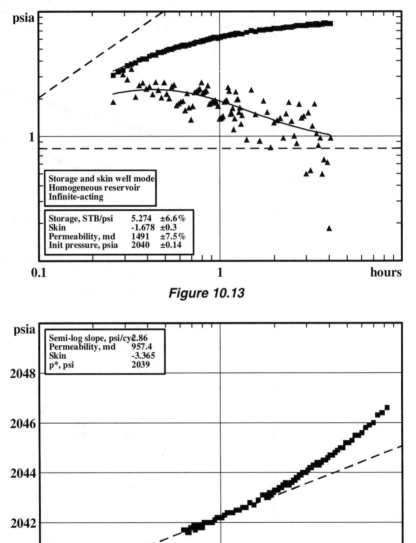

Figure 10.13

Figure 10.14

Test 6: Horizontal Well

There are some important issues to be noted with respect to the sensitivity to effective well length and vertical/horizontal permeability ratio. Fig. 10.15 shows a match to this buildup data using the known drilled length of 1700 feet, and a k_V/k_H ratio of 0.2 which is typical of the formation. The confidence intervals are acceptable, however the match to the derivative is unsatisfying, especially at later time. Changing the effective well length to 500 feet (Fig. 10.16) results in much more convincing match.

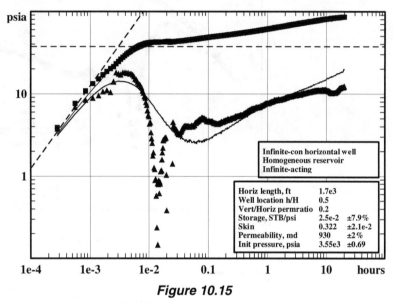

Figure 10.15

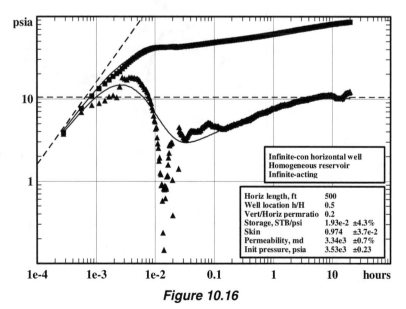

Figure 10.16

Although the effective well length is very important in this test, the solution is practically independent of the value of k_V/k_H ratio. Fig. 10.17 shows that if k_V/k_H ratio is allowed to float, the value is not determined with acceptable confidence interval. This uncertainty is emphasized in Fig. 10.18, which shows that the match is unchanged even with a k_V/k_H ratio of 0.1. It should be noted that the skin factor estimate changes with k_V/k_H ratio value, since the two are strongly correlated.

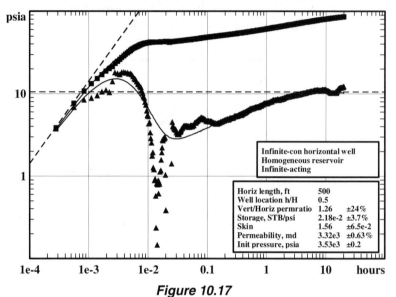

Figure 10.17

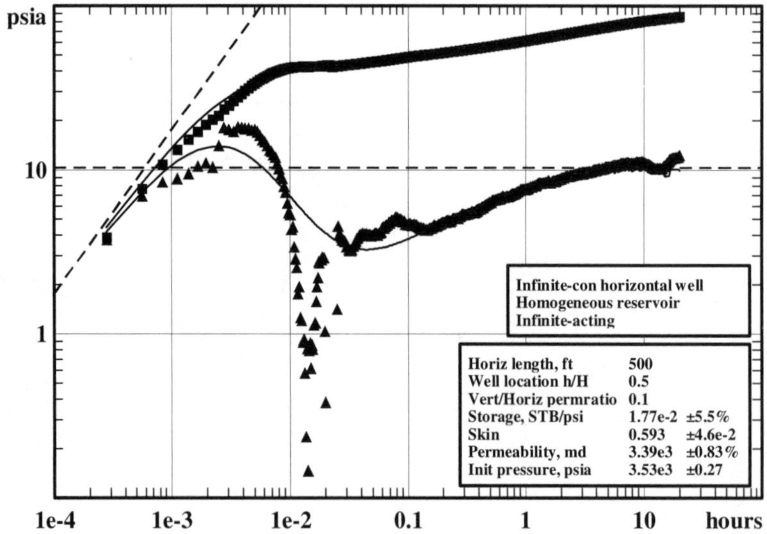

Figure 10.18

Test 7: Test from Mechanical Gauge

This test is difficult to interpret, because the data were poorly sampled, with insufficient data in the infinite-acting region. Fig. 10.19 shows a reasonable looking match, but poor confidence intervals. Fig. 10.20 shows a match with different values of *k* and *s*, but the matched curve looks almost the same as before. Resampling the data would be advisable here. Data recorded by hand from mechanical tools can sometimes be like this.

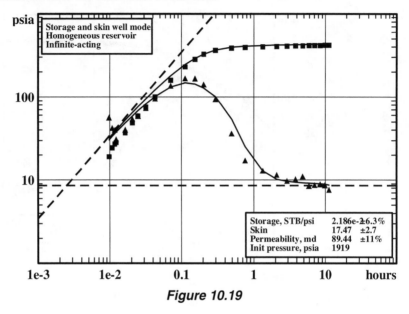

Figure 10.19

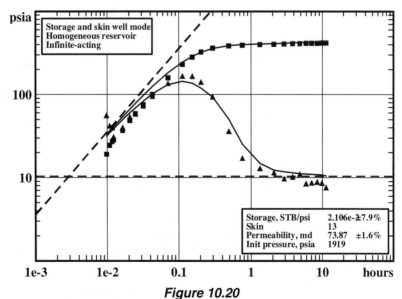

Figure 10.20

Test 8: Damaged Well

This is a conventional damaged well response. The latter part of the storage hump falls more sharply than it is supposed to (Fig. 10.21), which adversely affects the confidence that can be placed in the parameter estimates. Horner analysis (Fig. 10.22) produces lower k and s values, but generating the response with these estimates does not result in a good representation of the data. The well is clearly damaged, but the skin factor cannot be estimated precisely from this test.

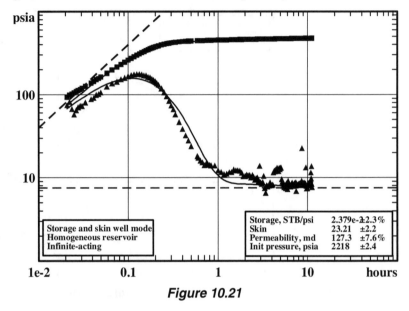

Figure 10.21

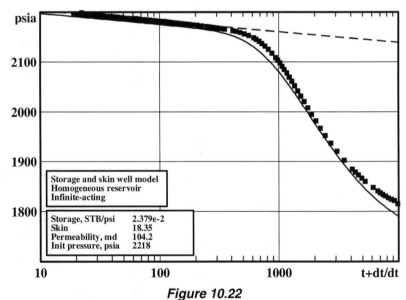

Figure 10.22

Test 9: Thermal Recovery Well

This test is a buildup test in a well in a "huff-puff" steam drive reservoir. The response shows the higher mobility inner zone, with the lower mobility zone outside it. The distance to the interface between the two zones can be estimated using the composite reservoir model.

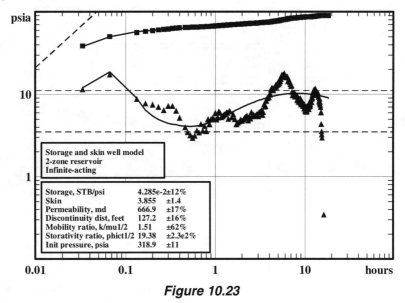

Figure 10.23

Test 10: Uninterpretable Test

Not every well test has an interpretation. There is no reservoir model that will explain this behavior. In all probability the pressure gauge has stopped working.

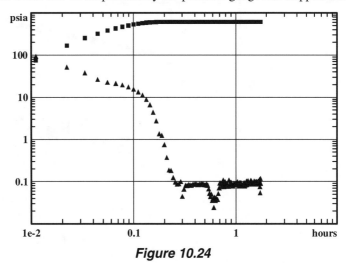

Figure 10.24

Test 11: Horizontal Well

This is a buildup test in a horizontal well. Experimenting with different values of vertical to horizontal permeability ratio reveals that this parameter is strongly correlated with the skin (which means it is not well determined). Conventional vertical-well techniques (such as Bourdet and Gringarten type curves or Horner analysis) do not apply here. The test does not reach late time radial flow, and early time radial flow is covered by wellbore storage effect.

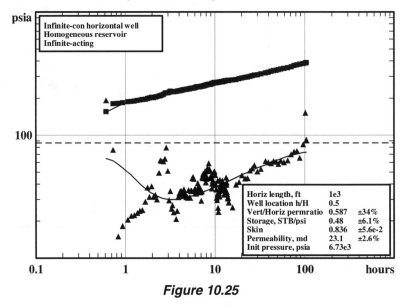

Figure 10.25

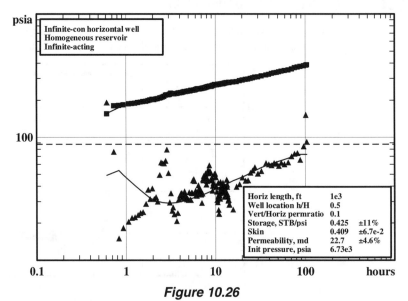

Figure 10.26

Test 12: Short Test

Compare and contrast the three buildup tests Test 12, Test 13 and Test 14. All have very short transients and do not properly reach the radial flow period. Of the three, Test 12 is the shortest from the point of view of interpretability. You can choose many different values of k and match to find a corresponding value of s -- the match looks about as good in every case. The test is far short of reaching radial flow behavior. Do not be mesmerized by the straight line on the semilog plot (Fig. 10.28) -- generating a transient using these values does not replicate the response.

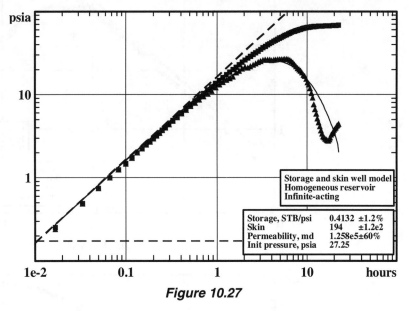

Figure 10.27

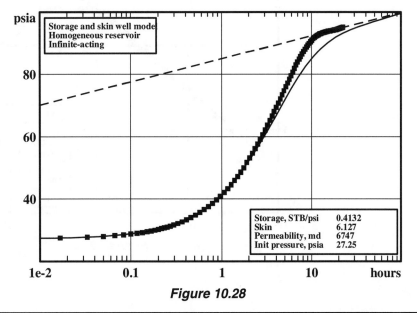

Figure 10.28

Test 13: Short Test

This is a buildup test, although the producing time was not recorded (a surprisingly common occurrence!). It can be analyzed as if it were a drawdown. The late time downward trend in the derivative could be due the first part of a double porosity transition, or to the buildup effect.

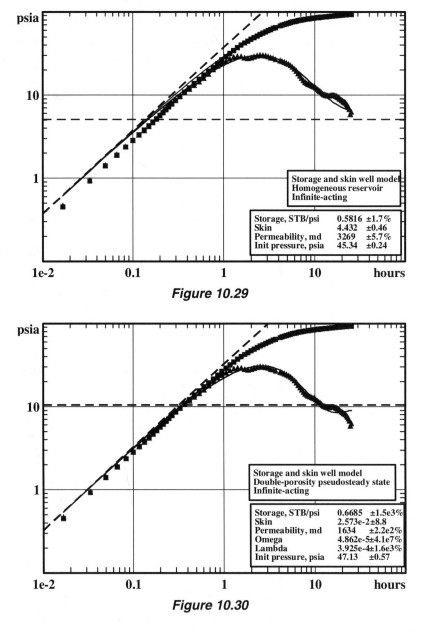

Figure 10.29

Figure 10.30

The dual porosity model provides a more convincing semilog straight line Fig. 10.32, but poorer confidence intervals. Although not as short as Test 12, this test is still too short for certain interpretation. One important point to note is that the two models shown here match the data almost perfectly on the semilog plot, and show only small variations from the data on the derivative plot. Hence we see that a good visual match to the data is not sufficient to guarantee a good answer -- the permeability estimates differ by a factor of three, and the skin factor estimates range from around zero to 4.4.

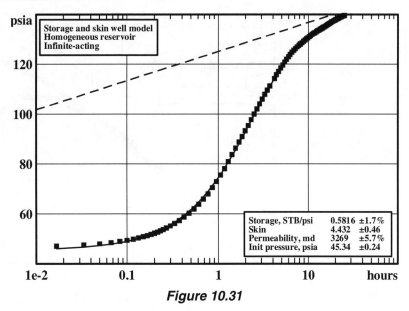

Storage and skin well model
Homogeneous reservoir
Infinite-acting

Storage, STB/psi	0.5816	±1.7%
Skin	4.432	±0.46
Permeability, md	3269	±5.7%
Init pressure, psia	45.34	±0.24

Figure 10.31

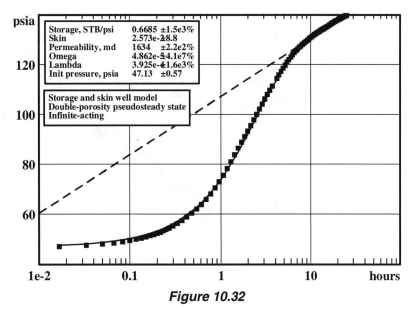

Storage, STB/psi	0.6685	±1.5e3%
Skin	2.573e-1	±8.8
Permeability, md	1634	±2.2e2%
Omega	4.862e-5	±4.1e7%
Lambda	3.925e-4	±1.6e3%
Init pressure, psia	47.13	±0.57

Storage and skin well model
Double-porosity pseudosteady state
Infinite-acting

Figure 10.32

Test 14: Short Test

The data can be matched to infinite-acting (Fig. 10.33), dual porosity (Fig. 10.34), closed boundary (Fig. 10.35) or to fault boundary (Fig. 10.36) models. Which is most appropriate? Actually none of them have confidence intervals within acceptable bounds, although the infinite acting model has the narrowest. It would be a mistake to choose a model on this basis however, since a model should have acceptable confidence intervals before it should be considered.

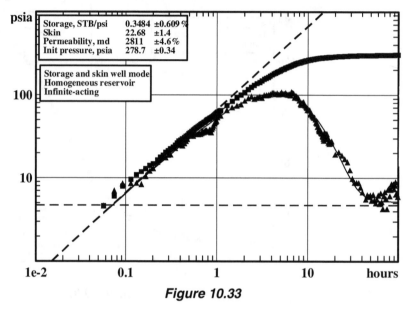

Figure 10.33

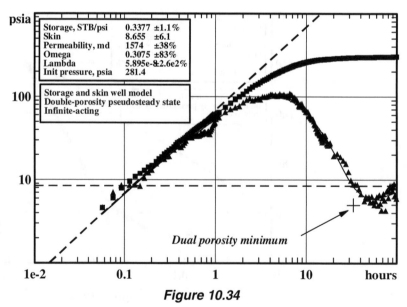

Figure 10.34

In an example such as this one, information external to the well test (such as geological data or other well performance) can be a guide to selecting the appropriate model. Test 14 appears to be longer than Test 13 (which is itself longer than Test 12), however is still too short for proper interpretation.

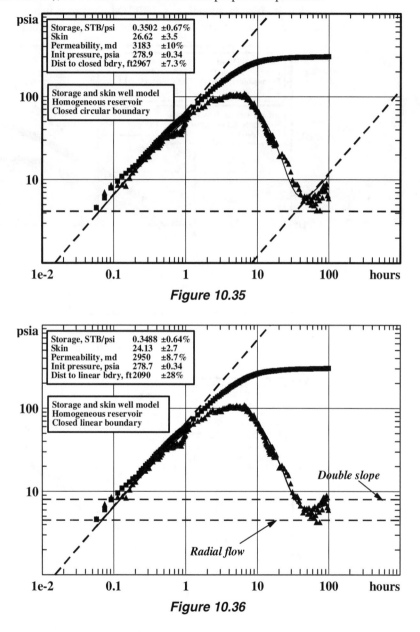

Storage, STB/psi	0.3502	±0.67%
Skin	26.62	±3.5
Permeability, md	3183	±10%
Init pressure, psia	278.9	±0.34
Dist to closed bdry, ft	2967	±7.3%

Storage and skin well model
Homogeneous reservoir
Closed circular boundary

Figure 10.35

Storage, STB/psi	0.3488	±0.64%
Skin	24.13	±2.7
Permeability, md	2950	±8.7%
Init pressure, psia	278.7	±0.34
Dist to linear bdry, ft	2090	±28%

Storage and skin well model
Homogeneous reservoir
Closed linear boundary

Double slope

Radial flow

Figure 10.36

The comparison between Tests 12, 13 and 14 emphasizes the need to ensure that a well test is sufficiently long to capture the essential behavior of the reservoir. It is difficult to overcome test design deficiencies during the interpretation.

Test 15: Hydraulically Fractured Well

This is a falloff test in a fractured well. The producing time (720 hours) is not very long compared to the test time (about 80 hours). The data fit well to an infinite conductivity fracture model (Fig. 10.37) although the match to the actual infinite conductivity fracture type curve is masked by wellbore storage effect (Fig. 10.38).

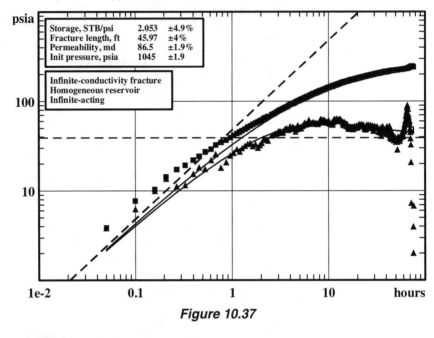

Figure 10.37

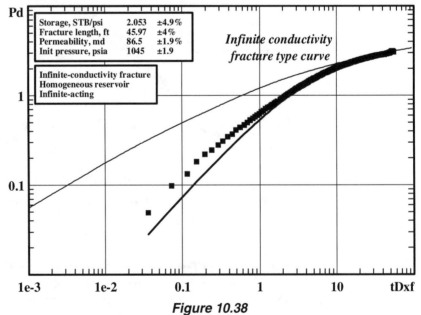

Figure 10.38

As with many fractured wells, the data can also be matched to a line source model with negative skin (Fig. 10.39). Notice that the permeability estimate is unchanged from the fracture model results (Figs. 10.37 and 10.38). Comparing the estimates of *k* and *s* from Horner analysis (Fig. 10.40) to those from regression, as in the case of positive skin the Horner analysis causes both parameters to be underestimated. The reason for this is that the data approach the Horner straight line from underneath (although this is on the top for this test since it is a falloff), hence the tendency to overestimate the slope of the line.

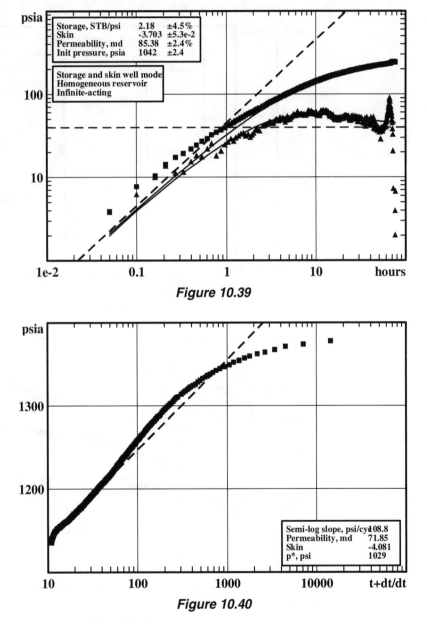

Figure 10.39

Figure 10.40

Test 16: Falloff Test with Prior Shut-In

This is a falloff test in which the well was closed about 24 hours before the eventual shut-in so that the tool could be lowered. Fig. 10.42 shows the analysis including the prior shut-in period, while the analysis in Fig. 10.43 ignores the earlier shut-in. Wiggles in the derivative are not relevant to the analysis (the prominent fluctuation at around 10 hours shows where the gauge was replaced). In general, derivative fluctuations less than half a log cycle in length are usually spurious. The data can also be interpreted using fracture models (Fig. 10.44).

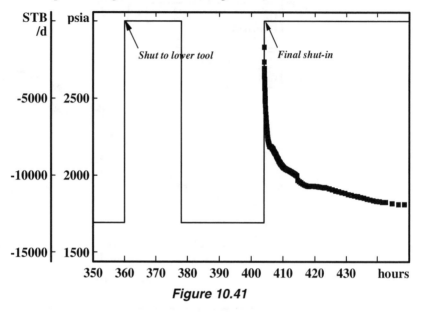

Figure 10.41

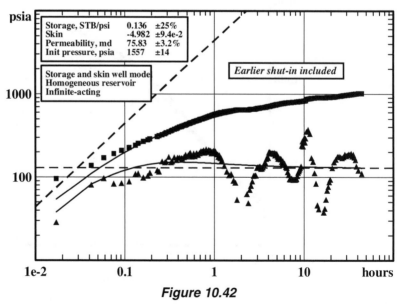

Figure 10.42

In this particular example, the prior shut-in does not have a very prominent effect on the resulting parameter estimates, however this is not always true. The length of the prior shut-in, as well as that of the flow period that followed, are both short relative to the length of the final shut-in (i.e. the test itself). However, since this well is highly stimulated, reservoir response can be seen from about 1 hour into the test. Hence the preceding shut-in and flow transients have minimal effect. Had the storage effect been larger, the influence would have been greater. In general, any preceding shut-in should be included in the flow history.

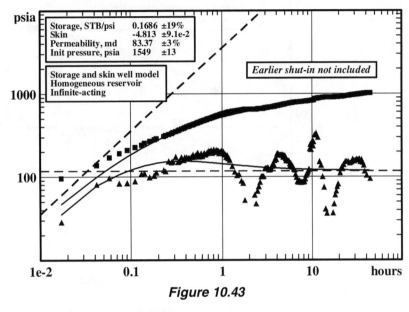

Figure 10.43

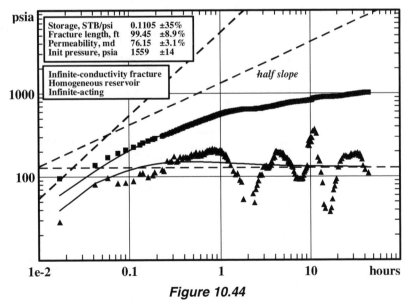

Figure 10.44

Test 17: Gas Well Buildup

This is a gas well buildup test. The results from analysis of this test are more or less the same either using or not using the pseudopressure approach (Figs. 10.45 and 10.46 respectively). This is because the pressures are relatively high throughout the test.

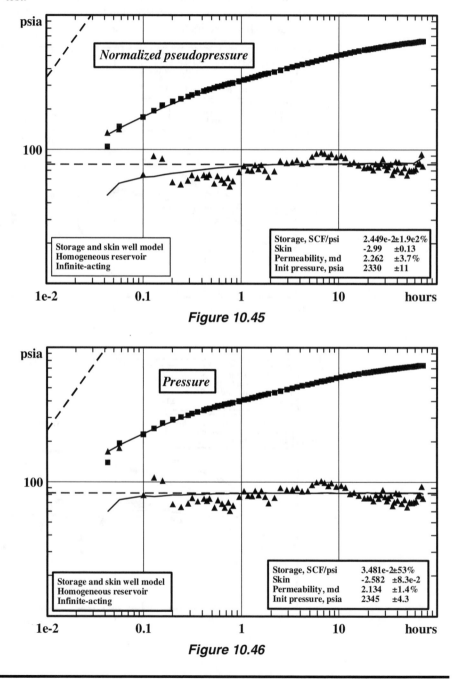

Figure 10.45

Figure 10.46

Test 18: Acidized Well

This is a buildup test in an acidized well. The data matches either to negative skin or to fractured well models. Since the wellbore radius r_w is 0.411 ft, a skin factor of -2.317 gives an effective wellbore radius $r_w e^{-s}$ of 4.17 ft, which exactly half the estimated fracture length. This is generally true -- a fracture acts like a cylindrical well with a radius half the fracture wing length.

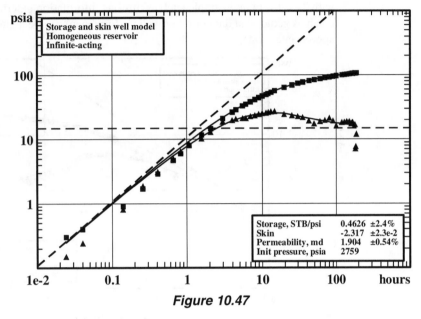

Figure 10.47

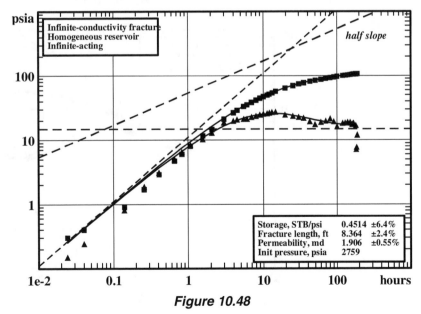

Figure 10.48

Test 19: Ambiguous Response

This test is a buildup test following a long period of production. The production period was long, and is also unknown, so the test is analyzed as a drawdown. The test can be matched using an infinite acting (Fig. 10.49), dual porosity (Fig. 10.50), closed outer boundary (Fig. 10.51) or fault boundary (Fig. 10.52) model. Confidence intervals are poor in the dual porosity and fault boundary cases. As with Test 14, in an example such as this one information external to the well test (such as geological data or other well performance) can be a guide to selecting the appropriate model.

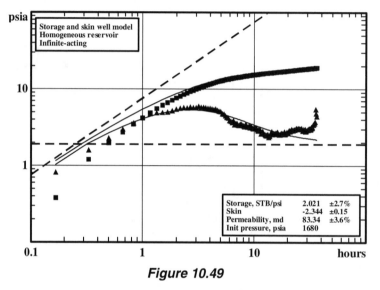

Figure 10.49

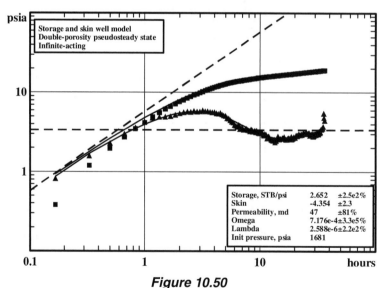

Figure 10.50

The two boundary responses both show radial flow lines that lie far below the data on the derivative. This usually increases the uncertainty in the estimated results. However, the closed boundary model (Fig. 10.51) shows better confidence intervals because the transition into pseudosteady state is modeled to occur during the time of the test. In the case of the fault boundary model (Fig. 10.52), the transition to double slope would be occurring later than the end of the measured data.

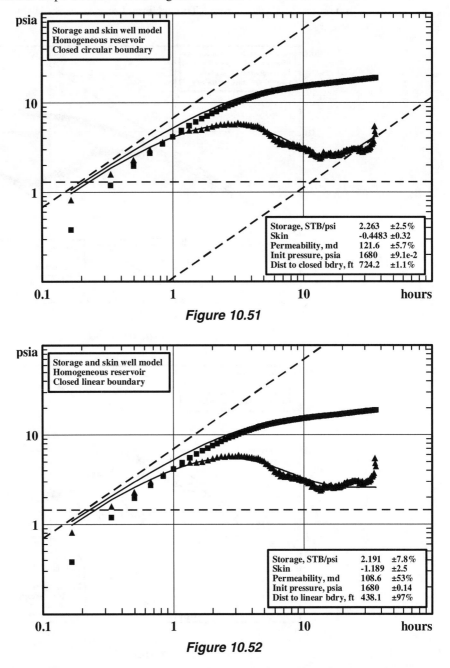

Figure 10.51

Figure 10.52

This example also illustrates what can happen when the boundary response starts before the 1½ log cycle transition is over, making it impossible to find the radial flow line. The data can be matched almost perfectly using skin values anywhere from -1 to 10 (Figs. 10.53 and 10.54). The confidence intervals on permeability look deceptively good in this case since skin has been held fixed to illustrate the point. Estimating the skin factor as well results in unacceptable confidence intervals for all parameters. This well requires a down hole shut-off for proper testing.

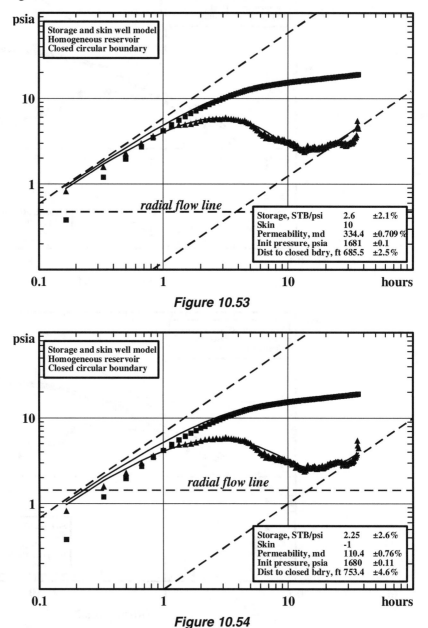

Figure 10.53

Figure 10.54

Test 20: Horizontal Well

This test reaches late time radial flow behavior, but unlike Test 6, there is also an indication of early time radial flow. The vertical to horizontal permeability ratio k_V/k_H, cannot be changed very much without affecting the match (a change from 0.2 to 0.1 is shown). There is some indication of hemiradial flow in this response, suggesting that the well may be closer than anticipated to the overburden or underburden.

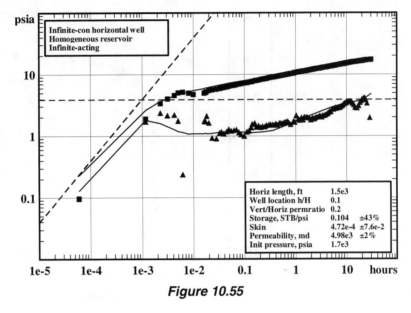

Horiz length, ft	1.5e3	
Well location h/H	0.1	
Vert/Horiz permratio	0.2	
Storage, STB/psi	0.104	±43%
Skin	4.72e-4	±7.6e-2
Permeability, md	4.98e3	±2%
Init pressure, psia	1.7e3	

Figure 10.55

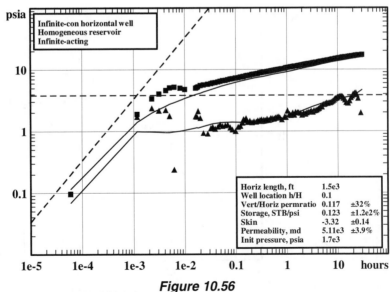

Horiz length, ft	1.5e3	
Well location h/H	0.1	
Vert/Horiz permratio	0.117	±32%
Storage, STB/psi	0.123	±1.2e2%
Skin	-3.32	±0.14
Permeability, md	5.11e3	±3.9%
Init pressure, psia	1.7e3	

Figure 10.56

Test 21: Downhole Flow Rate Measurement

This example was reported by Meunier, Wittmann and Stewart (1985), and the data may be found in Table 3-5 of the book by Sabet (1991). The data are from a buildup test in which the flow rate was measured down hole. Fig. 10.58 shows a rate-superposition plot, with a clear straight line behavior. Matching the storage and skin solution using all the measure flow rates, as in Fig. 10.59, provides parameter estimates with good confidence intervals.

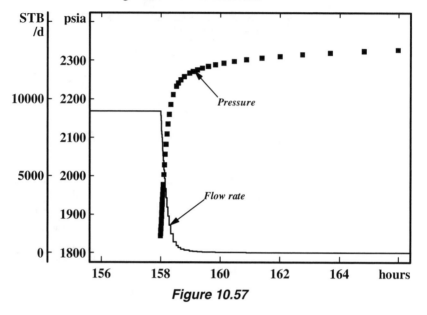

Figure 10.57

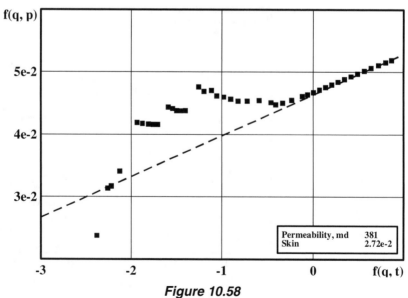

Figure 10.58

The estimate of wellbore storage coefficient in Fig. 10.59 is small, since the flow rate measurement all but eliminates the need to include this effect. If the flow rate data are ignored, the test may be interpreted in the usual manner (Fig. 10.60) resulting in a much larger estimate of wellbore storage coefficient, but the match to the data is not as good. Note that the derivative computation in Fig. 10.59 uses rate-superposition time, hence the disappearance of the storage region and hump.

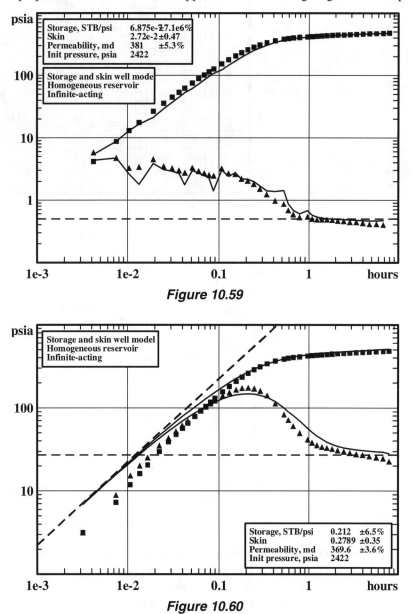

Figure 10.59

Figure 10.60

Combining the simultaneous measurements of pressure and flow rate using the Laplace pressure approach described Bourgeois and Horne (1993) allows us to see the underlying solution by deconvolution (Fig. 10.61).

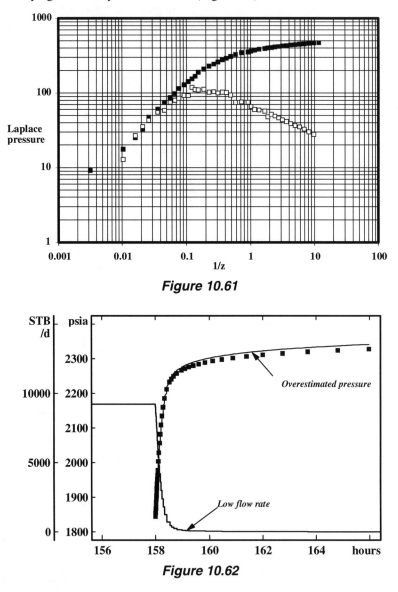

Figure 10.61

Figure 10.62

One difficulty with downhole measurements of flow rate is that mechanical (spinner) flow rate tools lose accuracy at low rates. This may cause the indicated rate to be less than actual, which gives a consequent error in the computed pressure (Fig. 10.62). It is usually not a good idea to try to extrapolate a smooth transition to zero rate -- a much better approach is to test the well without allowing it to stop flowing.

Test 22: Damaged Well Buildup

This data was reported by Bourdet, Whittle, Douglas and Pirard (1983), as listed in Table 4-3 of the book by Sabet (1991). The response is a classic example of a damaged well response, and is also useful to see the effect of the superposition of the preceding flow period (which was only 15 hours long in this case). A plot against real time (Fig. 10.63) shows the downward dipping derivative caused by buildup effects. Plotting against effective time (Fig. 10.64) removes this phenomenon.

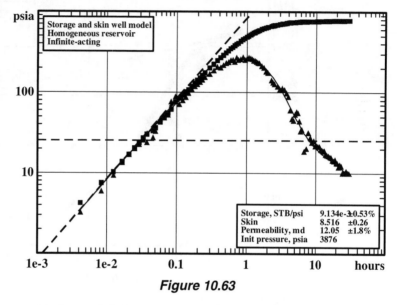

Figure 10.63

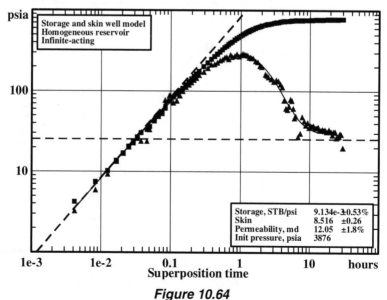

Figure 10.64

This test also provides a good example of the deceptiveness of the Horner plot. Since the data in a buildup test always approach semilog behavior from below, it is a human tendency to draw the Horner straight line too steeply (Fig. 10.65), resulting in estimates of permeability and skin factor that are too low and an estimate of final pressure that is too high. Plotting the Horner straight line in the correct position based on the match to the data (Fig. 10.66), the values of permeability and skin are much larger. The data only asymptotes to the straight line rather than following it for any period.

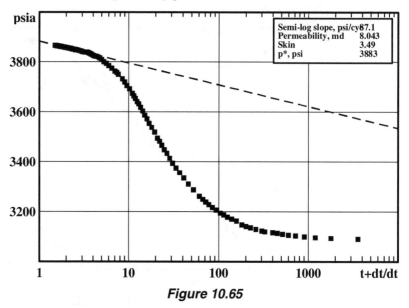

Semi-log slope, psi/cy	87.1
Permeability, md	8.043
Skin	3.49
p*, psi	3883

Figure 10.65

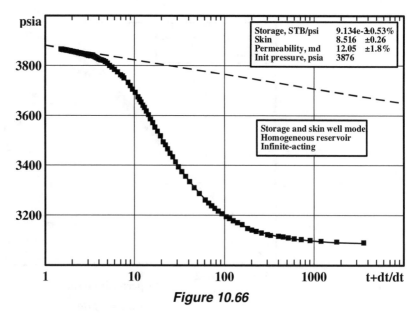

Storage, STB/psi	9.134e-3	±0.53%
Skin	8.516	±0.26
Permeability, md	12.05	±1.8%
Init pressure, psia	3876	

Storage and skin well model
Homogeneous reservoir
Infinite-acting

Figure 10.66

Test 23: Hydraulically Fractured Well

This drawdown test in a well with a hydraulic fracture was reported by Gringarten, Ramey and Raghavan (1975) and is listed as Table 4-5 in the book by Sabet (1991). The response shows the characteristic half slope with a factor of two separating the pressure and the derivative (Fig. 10.67). A good match is obtained to the infinite conductivity fracture model, and the data fits the finite conductivity fracture type curve at a large value of the fracture conductivity (Fig. 10.68).

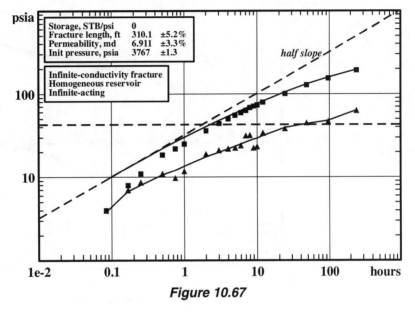

Figure 10.67

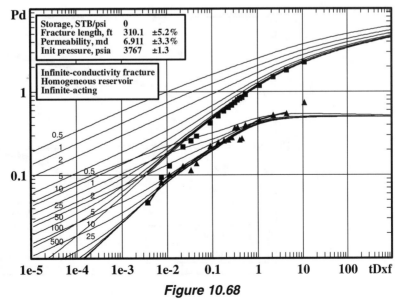

Figure 10.68

Test 24: Dual Porosity Response

This example of a dual porosity reservoir response was reported by Bourdet, et al. (1984) and the data are listed in Table 6-3 of the book by Sabet (1991). A derivative plot (Fig. 10.69) shows the minimum characteristic of dual porosity behavior. A better looking match is obtained if the transient interporosity flow model is used (Fig. 10.70), although in this case confidence intervals on the estimate of ω are poor.

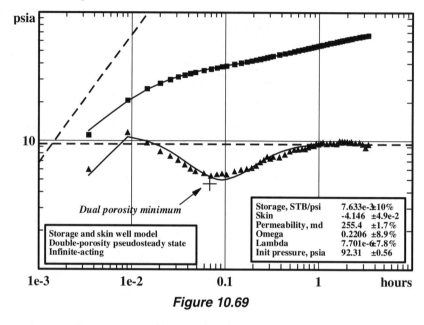

Figure 10.69

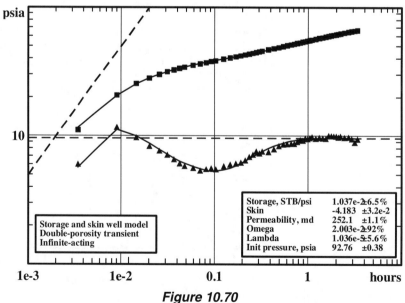

Figure 10.70

Test 25: Geothermal Well

This example is a buildup test in a geothermal well. Since geothermal wells usually produce from fractured volcanic rocks, they frequently show highly stimulated behavior (even though they are rarely stimulated). This example can be matched adequately using either a fracture model (Fig. 10.71) or a large negative skin (Fig. 10.72). The large wellbore storage coefficient is common in geothermal wells due to the large wellbore volume and the compressibility of the steam-water mixture in the wellbore.

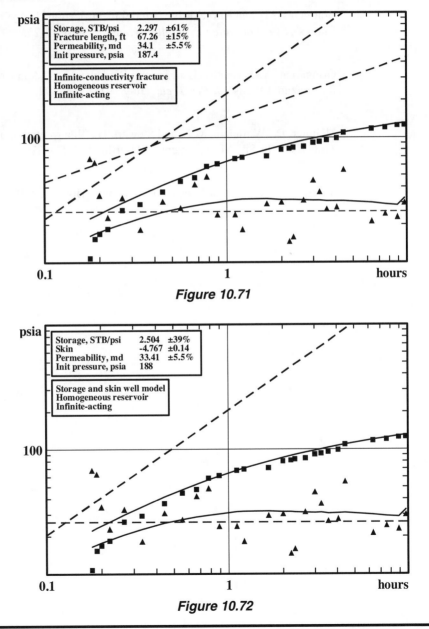

Figure 10.71

Figure 10.72

References

Bourdet, D., Alagoa, A., Ayoub, J.A., and Pirard, Y.M.: "New Type Curves Aid Analysis of Fissured Zone Well Tests", World Oil, (April 1984), 111-126.

Bourdet, D., Whittle, T.M., Douglas, A.A., and Pirard, Y-M.: "A New Set of Type Curves Simplifies Well Test Analysis", *World Oil*, (May 1983), 95-106.

Bourgeois, M., and Horne, R.N.: "Model Identification Using Laplace Space Type Curves", *SPE Formation Evaluation*, (March 1993), 17-25.

Gringarten, A.C., Ramey, H.J., Jr., and Raghavan, R.: "Applied Pressure Analysis for Fractured Wells", *J. Pet. Tech.*, (July 1975), 887-892.

Meunier, D., Wittmann, M.J., and Stewart, G.: "Interpretation of Pressure Buildup Tests Using In-Situ Measurement of Afterflow", *J. Pet. Tech.*, (Jan. 1985), 143-152.

Sabet, M.: *Well Test Analysis*, Gulf Publishing, Houston, (1991).

Index

1

1½ log cycle rule 25, 84

A

absolute open flow potential 96, 114
active well 6, 197
Agarwal equivalent time 57
ambiguity 1, 91, 168–69, 186
aquifer 2, 30, 34, 134
average reservoir pressure 3, 60, 123–24

B

bilinear flow 37, 45, 179
boundaries 30–34, 44, 46, 49, 134, 159, 167
  horizontal wells 134, 136
buildup test 5–6, 54, 57, 60, 96, 124, 175
  derivatives 76, 77, 191
  effect of boundaries 173
  effect of p_{wf} 175
  example 58, 170, 217, 220, 221, 223, 224, 225, 229, 230, 231, 232, 240, 241, 242, 246, 249, 253
  in dual porosity 187, 191, 192
  treating as drawdown 60

C

closed boundary 30, 45, 66, 174, 221, 234, 243
commingled flow 139
confidence intervals 89, 91–93, 101, 168, 185, 189

acceptable range 90
  example 155, 156, 160, 164, 167, 185, 189, 200, 217, 223, 225, 227, 234, 242, 244
consistent units 13
constant pressure boundary 30, 34, 48, 66, 75–77, 174
constant rate 5, 17, 46, 50–51, 65, 88, 94–95, 98–99, 115, 175
crossflow 140–43

D

damaged zone 13, 162
Darcy's Law 10
data preparation 84, 101
deconvolution 51, 98–99, 101, 248
deliverability 2, 96
derivative matching 191
derivative plot 44, 67, 76, 82–83, 151, 165–66, 170–72, 189
desuperposition 51, 77, 97–98–99, 101, 175, 248
  layered reservoirs 142
diagnostic plot (see derivative plot) 83, 217
differentiation interval 80–81
diffusion equation 10
dimensionless variables 10, 27, 113–14
double porosity (see dual porosity) 41–42
drainage area 32, 60, 114, 129, 173, 221
drawdown test 4, 123
  example 99, 145, 147, 176, 182, 199, 251
drill stem test (DST) 7, 102, 160–63
dual porosity 41–42, 82–83, 72, 76, 82–83, 41–42, 128, 82–83
  example 186, 188, 223, 233, 242, 252

E

effective wellbore radius 13, 14, 241
exponential integral 23, 48, 195, 197

F

falloff test 6
  example 236, 238
fault boundary 34, 169, 234, 242–43
finite conductivity fracture 36, 37, 70
  example 176, 179, 180
flow efficiency 14
fracture conductivity 38, 179, 181, 186, 251
fracture length 36, 177–79, 184, 186, 217, 241
fractured wells 36, 186, 237
fractures
  finite conductivity 36, 37, 70
  infinite conductivity 36–38, 71, 182, 236, 251
  uniform flux 36, 39–40

G

gas properties 207
gas wells 107
  example 199, 221, 240
geothermal wells 253

H

half slope 179, 182, 251
hemiradial flow 134–37, 245
heterogeneities 3
horizontal wells 133
  example 225, 230, 245
Horner false pressure 61
Horner plot 56, 58, 113, 174–75, 192, 250
Horner time 56–57, 76
hydraulic diffusivity 10

I

infinite acting flow 22–23, 22–23, 25, 32, 44, 57, 68, 83, 113, 128, 146, 184
infinite conductivity fracture 36–38, 71, 251
  example 182, 236
initial pressure 60, 87, 94, 154, 157, 191, 218
injection test 5–6
interference test 6
  example 130, 194
interporosity flow 42, 252
inverse problem 1

L

Laplace pressure 99
  example 248
layered reservoirs 139
least squares 88
line source 23, 195, 197
linear flow 37–39, 45, 220
  horizontal wells 134, 137

M

mathematical models 9
MBH plot 61
MDH plot (see semilog plot) 66, 187
multiphase flow 119, 214

N

non-Darcy skin 115
non-ideality 102
nonlinear regression 88, 90, 94–96, 99–100, 101–4
  ambiguity 169–70
  derivative matching 191
  example 156–57, 172–73, 175–76, 179, 184, 200, 202
  horizontal wells 138–39
  in dual porosity 189
  in interference tests 195–96, 198